LIGHTNING RIDGE

The Land of Black Opals

Ion Idriess

ETT IMPRINT
Exile Bay

This 23rd edition published by ETT Imprint, Exile Bay 2024.

First published by Angus & Robertson Publishers 1940.
Reprinted 1940 (two), 1942, 1944, 1947, 1948, 1950, 1951, 1953, 1956, 1959, 1961, 1968, 1973 (three), 1977, 1985.
Facsimilie edition published by Idriess Enterprises 2009.
First electronic edition by ETT Imprint 2017.
First published by ETT Imprint in 2020.
Published as an Imprint Classic in 2024.

© Idriess Enterprises Pty Ltd, 2017, 2020.

This book is copyright. Apart from any fair dealing for the purposes of private study, research, criticism or review, as permitted under the Copyright Act, no part may be reproduced by any process without written permission. Inquiries should be addressed to the publishers by mail or by email ettimprint@hotmail.com:
ETT IMPRINT
PO Box R1906
Royal Exchange NSW 1225 Australia

Cover: working at Lightning Ridge, 1910.
Cover and internal design by Tom Thompson.
Photographs on pages 56 and 74 appear courtesy
of the Lightning Ridge Historical Society.

ISBN 978-1-923205-11-6   (pbk)
ISBN 978-1-922384-97-3   (ebk)

CONTENTS

1. Skeletons in the Cupboard	5
2. Young Savages	9
3. The Silver City	13
4. Where the Camel Trains Go	18
5. In Wallaby Land	22
6. When Fever Comes	25
7. Back to the Bush	28
8. I Run Away	33
9. On the Track to the Ridge	38
10. The Loony Boundary-Rider	43
11. The Poison Carts	47
12. Riding the Boundary Fences	52
13. Lightning Ridge	57
14. Tom Peel	62
15. The Ratters	65
16. Trapping the Ratters	70
17. Black Opal	75
18. Stealing Grass	82
19. The Ruined Hut	88
20. The Stockwhip	92
21. The Sundowners	96
22. The Fright	101
23. The Monaro Cocky	104
24. Kaiser	110
25. The Count	114
26. The Opal-Buyers	120
27. The Old Man	127
28. On Opal	132
29. St Patrick's Day	137
30. Death	143
31. Deadman's Claim	148
32. The Trooper	154
33. Ghostly Company	162

AUTHOR'S NOTE

My publishers have suggested that I become autobiographical and tell something of my boyhood. "Begin at the beginning," they urged, "and give us a few glimpses of your early life. Your readers' curiosity is natural, and should be satisfied – cupboard skeletons and all."

Despite the reference to skeletons, I could hardly refuse. So, with cordial greetings to my readers, I offer the following.

I.L.I.

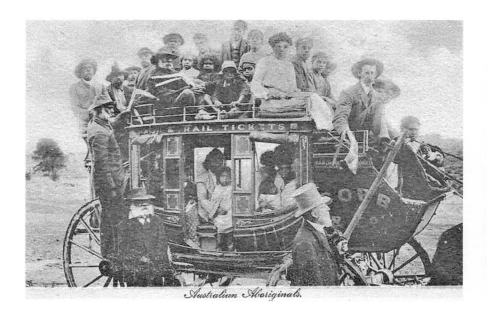

A packed Cobb & Co Coach, Central West, 1910 postcard.

1. SKELETONS IN THE CUPBOARD

"Dad, who were my ancestors?"

"What the blue blazes do you mean!" demanded dad. The family chatter ceased. (The meetings of the clans at the old folks' on week-ends are surprising.)

"I want some ancestors, dad. I have to write a book about them."

"Great Caesar!" Dad's a baldy, nuggety old boy with a leathery, weather-beaten face – an old sea-dog.

"Speak up, dad! surely I've had *some* ancestors!"

"Blow the man down!" he snorted. "Where do you think you came from – a hen-roost?"

"My publishers suggested something like that."

Dad appeared about to speak, changed his mind, reached for the smelly old pipe, and thoughtfully lit up. The family coughed. He gazed at me with suspicious yet serious eyes.

"You came," he declared, "from King Arthur's table."

"Be serious, dad. Who – were – my – ancestors?"

"King Arthur!" he roared. The family listened breathlessly, the womenfolk visioning coronets we'd somehow missed. "You are a descendant of King Arthur and the Knights of the Round Table."

"Heavens!"

"That's where you came from. You are descended from the King of Kings."

"Who was he?"

"Blow the man down! ... he was King Arthur."

"Oh!"

"We're descended from King Arthur. Every Welshman is."

"But I'm an Australian, dad. Anyway, how did these old ancestors put in their time?"

"Fighting, generally."

"What did they fight for?"

"Oh, mostly for the pleasure of a little blood-letting. Sometimes a matter of cattle. Maybe women, if there was nothing better to be got."

"H'm. Well, before King Arthur's day, who were our original ancestors?"

"God only knows."

"Well, then, who was the first big hit in the family?"

"They were *all* big hits, otherwise we wouldn't be here."

"I didn't mean with a club. Who was the first Idriess or Owen to make the welkin ring – to do something really big?"

"Young feller me lad," said dad soberly. "If I was to retail one hundredth part the great deeds of the Owen clans, there's not a library in the world big enough to hold the volumes."

"Sort of left our mark in history."

"The Welsh," said dad impressively, "made history."

"Were there any money-makers among them?"

"The Welsh," boomed dad, "did not raid for money, but to borrow cattle and cut one another's throats."

"Well, did any of the ancestors amass large herds of cattle?"

"They didn't live long enough," frowned dad. "No sooner had they built up a nice little herd than those thieving Llewellyns swarmed over the mountain and took them back."

"Couldn't hold their own like. No offence, dad. But surely some member of the clan amassed a fat herd."

"There was Griffith-ap-Idris," mused dad. "The Pixie helped him."

"The Pixie?"

"Yes. She came up out of the lake one moonlight night. Altogether lovely! Griffith caught her. She gave him his wish, a fine herd of fat cattle ... and a baby."

"H'm." I remembered the ancestral cupboard in time. "Well, dad, I never dreamed the blood of kings pulsed through the good old veins. But this fairy mixture is a bit heady."

"It's a solemn fact," declared dad.

"Tell us about an ancestor who did things, dad."

"There was Owen-Tudor," he replied thoughtfully.

"Who was he?"

"Lord bless my soul!" Dad sat bolt upright. "Do you mean to say you've never heard of Owen-Tudor?"

"I don't seem to recollect the bloke," I replied lamely.

"Owen-Tudor," explained dad with pained emphasis, "rode down from the Welsh hills, threw his weight about, and married Queen Catherine of England. He was the father of kings."

"Oh, I say, dad, one in the family might pass muster, but two would be fishy."

"Are you casting aspersions on our line of descent?"

"No, dad."

"You bally Australians — — he began.

"No offence, dad, but two kings take a lot of explaining away. Australians are a bit touchy about their ancestors."

"Australians," declared dad, "haven't got ancestors."

"H'm. Then we're descended from two lines of kings?"

"All Welshmen are."

"Didn't we grow any of the proletariat among the gang?"

"Just what do you mean?"

"Weren't there any common blokes? We couldn't all have been kings. Weren't there any bushrangers or pirates or something?"

"There was Morgan," remarked dad.

"What!"

"Morgan the pirate."

"Surely we're not harbouring a cut-throat like that in the cupboard!"

"Morgan," roared dad, "died in a feather bed as Morgan, His Majesty's governor of Jamaica."

"Sorry. I thought you said Morgan the pirate."

"Morgan *was* the pirate."

"But you said he was a governor."

"Lord bless my soul," declared dad helplessly, "don't you Australians understand the most elementary facts of history! Morgan was of the Morgan clan on Cader-Idris. He took to the sea to make his fortune. He was the great pirate Morgan, knighted by the king and made governor of Jamaica."

"I see. First, he slit throats, made 'em walk the plank, pinched their doubloons, and got away with it."

"Morgan!" declared dad, "was Morgan, governor of Jamaica."

"I see. Cut their throats, get caught and you swung. Cut their throats, don't get caught, and you are made a governor, and don't die a pirate."

"Your Australian reckoning is beyond me," declared dad.

"What have the old Idris ancestors to do with Morgan anyway?"

"Some of the clan sailed with Morgan."

"Ah. And what happened to the Idris boys among the crew?"

"They swung at the yard-arm."

"Dad, we're muddying the family escutcheon a bit, better glide back to the old ancestral tree. Who was your own grandfather anyway?"

"Owen-Idriess of course," snorted dad. "Fought under Nelson at Copenhagen and the Nile."

"Bravo, dad. Were there any witches in the family?"

"For God's sake what sort of a family do you want? I've given you King Arthur, a Pixie, Owen-Tudor, and a pirate – what *more* do you want!"

"The witch. All Welshmen have a witch in the ancestry, or ought to."

"If your damned Australian snobbery turns up its nose at your own ancestors then you can go and fish for them!" snorted dad. He reached for the paper and became an oyster.

Which brings us from the past to the present and dad, who got his discharge nobly or otherwise from H.M.S. *Nelson* and after sundry battlings met an exceedingly pretty girl fresh from an outback station. Dad won the engagement and the result was me.

Born in Sydney. The great event passed unnoticed. And yet mother believed me the most wonderful baby ever born. Dad wangled a temporary government job and thereafter most of his life was spent in travelling. I've been travelling ever since.

Earliest memories are of Tenterfield. There, in rompers and nearly three feet tall, I ran away from home, determined to see the world. A policeman with a big smile ended that premature escapade.

About the time of the disastrous bank smashes dad was offered a permanent job in Lismore, if he got there quickly. Hurriedly he loaded our scanty possessions on a bullock wagon opportunely starting out to Lawrence for stores. The bullocky shouted, the long whip curled up to fall with a crack like a pistol-shot. The bullocks tossed their horns, straining to a creak of chains, the big wheels squeaked, groaned, revolved, and dad started off, walking beside the bullocky. Dad turned round and waved. Mum and baby sister and I were to follow on later.

From the Lawrence, dad loaded up a sulky and set off on the rough trip down to the rivers after passing the Richmond Range, and was in nice time to drive right in amongst the big floods. He was the last man to dash across the old bridge into Lismore. The township was an island in a sea of creeping waters.

Scores of thousands of miles of wanderings in later years has forced one Australian at least to realize how capably the men from the British Isles took to the bush. They struck out and battled for their livelihood under conditions which would appal us to-day. The old hands were "as good men" as we.

2. YOUNG SAVAGES

Lismore in those days was a little bush township owning a dusty street down which the wild bush boys would gallop, to my delight and the swift action of the law. A very different Lismore to the town of to-day.

Our cottage was several miles out on the Nimbin road, overlooking the Richmond River. That cottage was the world. From its little attic windows I used to watch the punts coming and going away down river; sometimes even a ship with sails, or a steamer throbbing along. Or the goings on down at the blacks' camp. It was a thrilling night when old Topsy would come raging out of town from the Lock-up. Shouts and shrieks would send me running to the attic window, to watch the whirling firesticks as Topsy fought the camp.

A dense jungle scrub loomed behind the cottage. I loved that mighty forest with its dark, mysterious depths, its lianas like ships' hawsers drooping down from the leafy gloom, its cool silences in shadowed mystery stretching far in under the trees. With bated breath I'd venture into the edge of the scrub, listening to a bird calling from far away in, to the wee tap, tap, tap, of a woodpecker's bill, to the stealthy slither of snake or animal thing, or the steady boring of some giant beetle.

The fascinating bird-life used to make childhood days entrancing. The flash of brilliant wings through sunlight or gloom, the feathered beauty peering from foliage high above, the luring, ringing bird-calls. What were those wild birds saying? Ever sweetly challenging, urging timid footsteps to come away in and see. Those footsteps never ventured so far that they could not return; fairies must have guided them.

There was a thrill too in the echoes of the axes biting into the mighty scrub. There is something grim in the exultant, steely notes of the axe. The bite of the blow, the ringing echo, the jarring tear of bark and chip. Month after month that exultant song could be heard, culminating in a shuddering crash as a giant came down bringing smaller trees crumpling in a man-made tornado of destruction. For years it went on all along the northern rivers, all down that magnificent coast.

Alas! other parts of Australia have witnessed similar destruction. Millions of pounds worth of magnificent timber wasted in smoke. And the bird- and animal-life sacrificed with it.

A huge Moreton Bay fig-tree towered on Leycester Creek at the back of the cottage. To this massive tree strong men used to tie the logs with ropes, awaiting the turn of the tide to float the rafts down river. It was thrilling to watch the big logs rolling down stream, the men balancing and jumping from log to log as they fended the logs, guiding them away with long poles. Sometimes stream and logs would combine and surge together with a harsh, rumbling threat like distant thunder. The men would run and jump and leap from log to log shouting hoarsely, toiling with a fierce skill to break the jam. Breathlessly I would expect a man to slip and be squeezed to pulp between the rolling, grinding logs. But the men were too wonderful for that.

When all was going well they would sing as the raft went down river, going – where?

Childhood memories whisper of wild excitement; of the "cedar pirates", of rumours of trouble among rival gangs of timber-getters; particularly of one great fight when the rafts of the rival gangs became mixed up, or broke away in a fresh, or else the hawsers were deliberately cut. The fights and frolics of the rival gangs often enlivened the river.

School-days came. Teacher used to jog along in a sulky drawn by a cunning old pony, and collect the toddlers *en route*. Boorie was a tiny school of logs and shingles around which towered dense scrub. She was a nice school-teacher, fearful lest a toddler get bitten by a snake. Numerous snakes, blacks and greens and browns and yellows, carpet- and tree- and whip-snakes, used to crawl from the big scrub and attend school to the glee of the "big" boys. They were our heroes, especially when they killed a "black 'un". Each morning teacher would group us toddlers safely in the one clear patch from where we watched her and the elder boys armed with waddies scout around the school, poking into the bushes and amongst the grass. Then they would peer underneath the school; always several elder boys would bravely insist on crawling right underneath, to teacher's obvious anxiety. Then teacher would unlock the door and the elder boys would stamp in and rout out any crawlers before we toddlers were allowed inside.

The bush school-teachers of those days had not only snakes to contend with, but distances, bush-fires and flooded creeks, falling trees and bolting horses. There were accidents in lonely places; and sulky blacks sometimes called for courage and initiative at a moment's notice. The country also was just beginning to grow up. The cedar was nearly done, the pine would soon be done, the scrubs were fast being felled and burned off for sugar-land. During the cane-cutting season strangers appeared, seemingly roaming everywhere. Cheery days. The muscular men always had a passing joke for the little boy who solemnly watched them swinging cane-knives hour after hour.

The men used to load the cane in sledges and haul it down to the river, where they would carry it aboard the punts. Enormous loads they carried, jogging out from the river-bank along a plank to throw their load down into the big black punt. It was fascinating to watch them, especially when a man slipped and fell with a mighty splash.

Gangs of cedar-getters worked the forests of the New South Wales north coast. The huge logs were hauled out of the scrub by bullock teams or floated down the rivers to be pit-sawn into planks.

One day I repeated to mother what a man said when he fell into the river. She was shocked.

When the season was over the cane-cutters disappeared, and we of the bush were left to the bush. Those were the days before the dairying came. First, the explorers, then the pioneers, then the cedar- and pine-getters, then the cane-cutters, then the dairying. How often we've watched the bullocks toiling where now a glimmer of nickel and steel hums by!

Dad was transferred to Tamworth, a growing bush town then, almost a city now. School-days began in earnest here, but the outstanding memory is of the girl scholars and their technique in making love. They were original. I've passed through the school of experience since then but have never been approached so suggestively by the gentler sex. To impress their love for a boy they'd impale flies on a pin, then pass the grisly token along to him. A knowing nod, and the girl thrilled to the fact that the cavalier had accepted. The more flies impaled, the greater her passion. One girl was a Cleopatra, she could impale more than twenty flies on a pin and not one would squash. A squashed fly was scorned; no self-respecting boy would accept it. Each fly represented a kiss. The girl who could cover a pin with flies was "a beaut"!

I had my triumphs. Many a pin black with impaled flies I've deposited in the ink-well. He was a proud boy who filled his ink-well oftenest each week. This was known as "bogging the ink-well".

Another accomplishment of the girls was writing love-notes. These mostly consisted of "I love you", and innumerable crosses. I used to take mine home to mother.

We grew up young savages. The bush was our adventure place, the elder boys our heroes who could find and kill snakes, even shoot wallabies, swim in the deepest waterholes, climb the highest trees. A few bad boys had wagged it from school, several had even thrown an ink-well at teacher. These latter we looked up to with awe, certain they would become bushrangers.

On Saturdays we'd meet "down at the Common" with our dogs. Then "the mob" would "take to the bush".

Every bird, wallaby, bandicoot, snake, scorpion, spider, centipede, rat, anything that could move quickly enough, would waste no time in escaping our eyes or the eager noses of the dogs. Under a sheet of bark we'd catch a big centipede, a "beaut" spider, or some other repulsive, vicious fanged brute. Then an elder boy would manoeuvre it into a matchbox until we'd found another as big and hideous. We'd tickle the fiends with straws until they became fighting mad, then tip them together. A fight to the death followed.

With the passing of time, dad was transferred to Broken Hill, far away in New South Wales' south-west. It nearly broke my heart to part with Johnny Allsop. Johnny was my mate. Alas, I was never to see Johnny again. He was killed in France.

3. THE SILVER CITY

Broken Hill was a roar of sound. Day and night the great mills were thundering, crushing their millions of tons of stone as the years rolled by. This was the Silver City, its great black mines silhouetted along the skyline on their ridge of silver and lead. Great men affect the destiny of nations. And these great mines were to affect the destiny of Australia.

Commencing school life at Broken Hill was misery to a shy bush boy. Here was a big fence, echoing to the shriek of city barbarians. Here frowned a big brick school, an army it seemed, of masters. Surrounding it were streets of houses under a fierce sun with a red duststorm advancing to overwhelm the town. Those pillars of dust would spread to the sky, then envelop the Line of Lode, while the howl of the wind vied against the roar of the stampers. Through the blood-red haze would be ghostly visible the huge black poppet heads, the roaring of the mills dying away to suddenly brazen forth again to a wavering of the wind. Then the storm would howl down on the town where frantic housewives were closing windows and doors and jamming carpets and clothes into cracks in a vain effort to keep out the dust. Summer heat and dust meant misery to many a woman. Occasionally a "Black Friday" would blot out the town; every house close shut, every shop shut, the lights inside invisible, the town smothered under the swirling black cloud of the Dust Fiend.

Several schools in Broken Hill had the doubtful honour of attending to my further education.

Proving too dull a scholar to be proud of, the teachers allowed routine to follow its course and called me up for two, four, or six cuts as occasion required. One teacher nicknamed me "Old Stick in the Mud" ... And Old Stick in the Mud, bent over his desk, would determine to stick in the mud. The only subjects I liked were history, geography, and the writing of short essays. But these subjects were only slurred over for half an hour each week. School life was almost entirely devoted to arithmetic, grammar, algebra, Euclid and Latin. Old Stick in the Mud knew nothing about them when he left school, and knows as much to-day. I still remember Euclid, because teacher told us there was a "donkey's bridge" in it, and I was the donkey.

The young of the human species are notoriously blood-thirsty. My first air-gun was a joy which, alas, left a legacy of ghosts, the unfortunate birds that died by it.

Saturdays and holidays were bits of paradise – there was no school.

So we hunted.

A harsh, sombre district this, very different to the bushland streams and forests of the country nearer the coast. Surrounding this town were bare flats and sterile hills. When possible to sneak away from being nursemaid to the sisters, off I went with the air-gun. It fired one pellet of shot a time, just capable, unfortunately, of killing a small bird. I would stalk the little brown ground larks on the flats with the bloodthirsty intensity of a tiger-hunter. Crawling over the rough brown earth for hours, making every coarse tuft of grass afford as much cover as a bush, "freezing" when flushing one suspicious little bird. He would hop on a stone the colour of himself and, quite motionless, gaze around seeking the danger instinct warned him of. I would not move an eyelid; so presently he would turn his tiny head as his beady eyes sought the hidden menace elsewhere. Inch by inch I would crawl closer. His head would turn again, but the stalker was part of the earth. Presently, very uneasy, he would hop to earth and run, his head low, his wee body swaying as he dodged away among the scanty grass-tufts and little stones. He was brown as the parched ground, and he clung close to the harsh earth which had reared him.

Subconsciously I was learning how all things of the wild so naturally take cover; was intuitively realizing that with all things life is a ceaseless fight against death. That no matter how small the living thing is, it has its joys and sorrows within its own little world, in which day and night life is one alert struggle to live. This wee lark whose life I was seeking had his place in a definite world of his own, among his friends and rival larks, learning to know and get the best out of his own private and communal feeding-grounds, and out of life. He had possibly been doing nothing in particular, or else was eagerly going a-love making; possibly was worried over family affairs while outwitting his enemies the keen-eyed hawk, the lurking snake, the prowling goanna, the terrible cat, the awful small boy. This lark's world, although it contained awful perils, held everything in the world that he wanted. He was free and capable, and utterly independent. There was only one faint, wispy, invisible thing given him, the one thing he needed, the one thing he could not do without – life. And I was trying to take it from him.

To this lark each coarse grass-tuft was individually known, not only for its crop of grass-seed as food but for the shelter it afforded him – how often have I watched a little fellow crouching under a tuft, invisible to everything but the terrible eyes of a boy. He was the colour of the earth and the sun-scorched tuft and the shadow it threw over his grey-brown body. Between the tufts would be bare red earth and across this danger zone he would flit at his greatest speed. Years later, how often I was to do the same thing, every nerve a live wire, heart and brain a madly throbbing thing, running as I never dreamed it possible to run, racing across a bare patch under a hail of machine-gun bullets!

The little brown stones were as familiar to the lark as cross streets and houses to us. He would wheel around one group of stones and thus momentarily hidden would dart off at a tangent seeking shelter ahead. But immediately his head had turned in flight I would be running on tiptoes in pursuit, and as soon as he vanished I would be down on hands and knees hurrying closer; then flat out on chest and toes draw yet closer, but cautiously now. When near I would draw forward noiselessly and with a movement capable of instant stillness; many a time a hunted lark has hopped on a stone and stared straight at me, stared long and hard. But the utter movelessness baffled him. Sooner or later his uneasy little head would peer away to seek the danger he could not understand. And his movement gave the chance to aim that terrible little gun – when he looked away. If he'd seen the faintest movement he would have flown. But these little ground things if seeing no movement, though feeling themselves in danger, trust for safety to their legs and camouflage among the grass and stones. For while not certain what the actual danger really is, they know, should they fly, it might be straight up into the talons of a hawk.

At home that gun was the terror of the mice, our neighbours' as well as ours. In the corners of the rooms, under the verandas, and the sides of the houses they had burrowed their wee holes. I would sit for hours if need be, utterly still, within range of a mouse-hole. The tiniest peak of the daintiest little nose would appear, the wee bristles twitching as he smelt this way then that for danger, or bread-crumbs. His wee head was there, his little forepaws, his tiny, alert little ears, his bright black beads of eyes. Then half his little grey body. The gun was ready levelled. His pretty head would turn this way, then that, before he ventured out to forage. But the gun would "puff" and the mouse would be kicking there half in, half out of his doorway.

To hit half a mouse, to stalk a ground lark out on the open flat and unerringly hit it with one pellet of shot, takes doing. Strange that these poor little victims, and the many things subconsciously learned and stored in the back of some brain-cell should have been actually the first training of a sniper! Life often appears incomprehensible. Surely Fate did not deliberately will that from the shooting of mice and larks a small boy should develop to the shooting of men. And yet, I was (among other duties) a sniper in the war. I would not be here now but for the lessons learned by studying the self-preservation instincts of birds. The poor little larks might forgive me if in the Land of Spirit Birds they could know that, when the time came to meet a man as good as myself and as well armed, I did not flinch.

Money was woefully scarce, but nature gave us youngsters one harvest nearly every year, the Sturt's desert pea season. The young plant is almost indistinguishable from the dull grey of the saltbush and bluebush and drabness with which nature invariably clothes her harsh places. We keen-eyed youngsters would roam many miles seeking these scarce plants; it was a joy to find one. Carefully transplanted into jam-tins, then hawked among the suburbs, they readily brought "a shillun for a big un missus! sprat for a little un". The plants when successfully grown would run riot and blossom into a blaze of scarlet and black. One of the world's most beautiful flowers, growing in only a very few arid and desert places, it is an everlasting glory carrying on the name of the great explorer.

One year we cleaned up a fortune; though it didn't last long. We sold sparrows as canaries, a bob a time. Ginger got the idea, I think, but it was young Morrison the tailor's son who thought there "might be money in it". We caught the sparrows in brick traps baited with bread-crumbs – the sparrows kicked up a great fuss.

The first experiment failed. Joe souvenired a tube of his sister's yellow paint and we lathered the sparrow with it, then put him out in the sun to dry. But it was oil paint. The miserable little chap was a sight with his feathers glued to his tail; he looked sea-sick. So Bill kept it for his elder brother who was courting a girl, and earned a shilling. This bought our first water-colours but there were only two yellow cakes in the box. We mixed them into a thick fluid and forced a sparrow to take a bath. What he swallowed was his own fault. But *he* didn't look much. Even when we'd used up all the paint on him he looked a dejected half-breed with washy brownish feathers showing among the yellow.

Doubtfully Bill and Ginger hawked him as a "young bird just comin' on, missus. Yes, he does look a bit sick – they all do when they're caged for the first time. No, he ain't goin' to die, mum, he'll get over it. Yes, mum, he's a cock bird, see the way he holds his tail! Thanks, mum."

To our delight the innocent bird lover bought him for a shilling. Enthusiastically we scouted the business-quarter seeking a shop that would sell us "all yeller paints", realizing there'd be nothing in it if it cost us a bob every time we painted a bird. The next attempt proved a better job, the "canary" retained some life and was a lot "yellerer", while costs had been cut down considerably. Snowy, whose ambition was artistic, put the finishing touches on him. He tried to paint its eyelids but the bird objected too vigorously.

This fraud sold readily. We were kept busy making cages, trapping sparrows, and painting them. Our salesmen developed a real technique:

"Only a shillun, missus. Look, he's a real cock bird, a young un. He's a great whistler; starts up every mornin' and keeps it up all day. There ain't no other bird like him in all the Hill. See his tongue! A whistler's tongue, mum, an' his tail is a beaut! You can tell by it he's goin' to grow yellerer an' yellerer! A pure bred. A champion whistler, mum, an' only a bob! Thanks, mum."

We sold scores and scores of them; made it a welter while the going was good, avoiding any street where we'd made sales.

In every respectable canary's cage there is a little drinking-jar and bath. The customers received their first shock on hearing their whistlers agitatedly chirping, to the amazed accompaniment of noisy sparrows outside the veranda. But when each morning the "whistler" had his bath and grew more disreputable and browner, and all doubt was washed away, the household roared with laughter at mum. If all those mums could only have caught "those little devils!"

It was a scared little crowd of conspirators who crept into the shed one evening: "The Bobbies are after us!" The bubble had burst. Hurriedly we dismantled the brick traps, dumped the wire for cage-making, buried old paint boxes, and otherwise hid all incriminating evidence.

Mother presented the family with another sister in Broken Hill and embarrassed Ion became a temporary nursemaid again. At last there came a never-to-be-forgotten Christmas holiday. I announced that I had taken a job with Mr Williams the grocer, driving his grocer's cart.

Poor mother was horrified. For some mysterious reason she imagined that her white-haired boy was destined for high hats and frock-coats and all manner of queer destinies that mothers dream of for their sons. And now the prodigy was driving the grocer's cart, at seven shillings and sixpence per week. Dad looked glum, as dads generally do. But it was agreed on that I could hold the job until school started again.

So each afternoon saw me jogging down Oxide Street with the cart loaded with groceries. One day the cart was piled full of eggs when the old horse bolted, all out, down Oxide Street with me hanging desperately back on the reins. Telegraph-poles sped by as we bounced from the gutter to miss a buggy by inches and a rooster by a feather. To jump out was a dangerous temptation. But even death seemed preferable to the disgrace of lumps of horse, cart, and eggs strewn all over the landscape. At last the old horse pulled up, blowing and puffing ... We had not smashed one solitary egg.

Christmas holidays were over all too soon. But the writing was on the wall.

4. WHERE THE CAMEL TRAINS GO

"Boy wanted. Apply Medical Hall, Argent Street."

My new responsibility was to sweep the shop, wash the medicine bottles, and not listen to the doctors' yarns. The gay sparks amongst the doctors and chemists there educated me in the mixing of ointments and pills and, incidentally though not officially, in the preparation of a certain powder which reacts, when secreted in a private utensil to which a liquid may confidently be expected to be added, with an embarrassing hissing and fizzing most disconcerting to the person concerned. The Medical Hall was a large and busy establishment, and the occasional jokes of its staff only helped to increase the efficiency of its chemists and doctors, some of whom have made a big name in the Hill.

The world suddenly seemed a very bright place. At weekends we lads would go a-roving, perhaps out to the dusty camel-camps to meet the bearded 'Ghans coming in with their long lines of camels, or to watch them starting out into the heat haze. Weirdly they would dissolve like shadow giants into mirage, bound for the Darling, for Milparinka, Mount Browne, White Cliffs, Wilcannia, the Cooper, Birdsville, Bedourie, the Lake Eyre sandhills, to distant stations in south-western Queensland, to frontier stations on the edge of the Simpson Desert. Surprising trips those camel-trains made across the saltbush and mulga, stretching from the Hill out into the arid lands distant from a train like the tentacles of an octopus. One caravan would be slowly plodding to cross the wastes towards Mount Hopeless and bring news and hungrily awaited stores to the men, and, yes, women and children battling there. In later years I met those women and saw the hunger in their eyes as they stared across to the horizon seeking the ships of the desert that would bring not only stores but clothes, and papers with pictures of women and stories of women's life in a world far distant.

Another caravan would branch off more to the north-east and, crossing the great lignum swamps, carry on through bushland across the Queensland border and on to the Paroo and Eulo, the country of the "Eulo Queen". A character in backblocks development was the Eulo Queen.

We knew many of the 'Ghans leaders. Mahomet Ali with the fierce eyes and big black beard, Abdul Khan and Ghenghis Khan, Abdul Futabulla and old Valait Shah, the broken-nosed priest. He brought dad a length of beautiful green silk and implored him to wrap it around the holy Koran upon which those in trouble took the oath in the Broken Hill Court House.

Strange tales the 'Ghans told us, squatting there, smoking their bubble-bubbles in the smell and shadow of their lumbering, snarling beasts. Ambush and robbery and tribal skirmish among the crags of the Afghanistan hills, border raids and stealthy deeds by night. Australian stories too of camel-trains lost in howling dust-storms in the awful Lake Eyre country, men and camels smothering in an inferno of dust, raving mad with thirst. Stories of fights with wild blacks, and of vengeance against one another. For among these 'Ghans were some who had brought personal vendettas all the way from India, and they fought them out among the valleyed sandhills fringing the South Australian border.

Many years later I was travelling across those red, pine-tree crested sandhills; harsh and desolate under a brazen sky by day, weirdly lovely by night, terribly lonely always. A raging dust-storm uncovered the skeleton of a man – many skeletons are uncovered that way. This one had a little round hole bored neatly through the centre of the forehead. And the bullet was inside. Beside the bones lay an Afghan knife, wonderfully preserved.

Boyhood tales of the 'Ghans flashed through my mind. Here lay possibly one of those very men; evidence of a vendetta, of the vengeance that never dies.

From Queensland, great mobs of cattle would come bellowing across the distant border *en route* to the hungry Hill. We'd meet them and yarn with Kidman's drovers. Listen to stories of stampedes and of blacks, of wild country a thousand miles to the north-west, of dead men's bones in lonely places, of craftily laid traps for cattle-duffers, of wild chases after horse-thieves, of epic droving treks down the famous Birdsville track.

Sid Kidman, the struggling drover-stockdealer, was then laying the foundation of a career which was to make him the Australian Cattle King of all time. Who would have dreamed that the boy listening to the stories of his drovers was one day to write his life story!

Time rolled on and I joined the boy staff in the Assay Office of the Big Mine, the Broken Hill Proprietary Coy Ltd. An interesting job; but the demand for assayers is limited.

We lads under the stern eye of Johnny Crafter the Samples Room boss had to crush and sieve the ore samples, crack coke and build up the furnace fires, and in general, be handy lads to the assayers.

Work in the Fire Room intrigued us. We wondered at the chemical mixtures which in their correct proportions under the fierce heat of the furnaces melted the hardest rock, and separated the worthless material from the shiny bead of pure metal.

In the Wet Room the assayers toiled with their delicate balances and beakers and acids which bit into the crushed ore and dissolved the metals and worthless gangue into a liquid, revealing to a decimal point the metals contained. That was where we first breathed gas, whiffs of those poisonous, heavy, hideous, yellowish-brownish-greenish chlorines and bromines which in certain assays escaped from beaker, oven, or flue. An assayer generally protected his lungs by wearing a wet handkerchief tied over his nostrils and throat. Thus we gazed at our first gas-mask, never dreaming that a time was coming when we must face such terrible fumes hurled at us in concentrated form; never dreaming that man could ever be so vile.

This ceaseless assaying was routine work done for the guidance of the management; to ascertain daily not only the ore tonnage raised but also the silver, lead, zinc, and gold content in those hundreds and thousands of tons which the miners raised year by year.

Now and again, one of us boys would get a real thrill, accompany an assayer below to get a sample of gas. Far down in that wonderful mine great fires were smouldering in places. For years they had to be fought and carefully watched lest they break from control and flood the galleries with fumes and disaster. Holding up the miles of underground drives and workings was a great forest of oregon. In places, this timber was smouldering. The danger areas were walled off, doused with continuous curtains of water, smothered under tailings, smothered with fire-quenching gas. But still, away down there in the bowels of the earth the fires smouldered on and periodically would burst into flame and fill the workings with poisonous gas.

The creeping of that deadly, invisible gas was a nightmare for all concerned. When a fire broke out the Big Mine broke into feverish scenes of organized activity. Wonderfully cool organizing, fine bravery, fierce work. A grim fight in the bowels of the earth with half-choked flames illumining the tunnels, misty fumes and invisible gas creeping through the drives, great logs splintering to burst into flame under pressure of thousands of tons of rock creaking and groaning as the supporting timbers snap like breaking trees and smoulder, then burst into flame.

We boys little realized we were working in an office fast helping to build up Australian industries. Sometimes a boy would be detailed off, for laboratory work, to crush the special samples for those mysterious "Big Men" the experimenters, the metallurgical chemists concentrating on solving deep secrets of nature. Forcing her to part commercially with metals and minerals she had hidden in minute form deep in valueless masses of rock or tailings. Those men toiled through years of experiments, despite thousands of heart-breaking failures. The story of their success is a romance engraved deep in the developmental history of Australia. They made the Silver City. Made possible the Newcastle Steelworks. And a hundred new industries which are helping to bring Australia to nationhood to-day.

Any boy detailed to crush samples for the Big Men was a proud boy indeed.

The Big Mine at that time was under the management of G. D. Delpratt. And experimenting at the Assay Office were such men as E. J. Horwood, Leslie Bradford, A. D. Carmichael, E. T. Henderson and other metallurgical chemists long since famous the world over.

At that time Carmichael had just about solved the problem of extracting zinc from waste tailings; Bradford, Horwood and Henderson were experimenting with flotation processes to separate various sulphides. Bradford discovered the use of retarding and accelerating re-agents in the Flotation Process, now used all over the world. The work of these men was to turn mountains of waste tailings into gold. And to do immeasurably more than that. Their discoveries were to produce in many fields work and money for mankind.

To perfect ourselves to take our place on the assay staff and, if we had it in us, even among the metallurgical chemists, the Broken Hill Proprietary encouraged us boys to attend courses on assaying, chemistry and metallurgy at the Technical College, the Broken Hill School of Mines. So, when the day's work was done, we would hurry home, enjoy a hasty wash and meal, then return to town to the big night school. Solid work for growing lads! But it gave us the opportunity to make good.

5. IN WALLABY LAND

Week-ends were happy times. Four of us would hurry home from work, rush for the old spring-cart or buckboard, throw in our guns and blankets, then heigh-ho for the hills: Tommy Abbott with his cheery laugh, Harry Mudge with his quiet smile, Bill Gaynor with his joke and song. We'd jog along all afternoon and far into the night. By the small hours we'd be thirty miles out in the midst of sombre, rocky hills scantily clothed in Dead Finish, wattle and quandong. Unharness the old nag, water and feed him, then boil the billy under the stars. Water was imperative. In that arid country water had meant the difference between life and death to the early prospectors. We learned the secrets of rock-holes, of soakages, and the flights of birds to water.

In the driest creek there was invariably a soakage somewhere, water lying below the hot sand. The wallabies used to dig for a drink, "smelling out" the water then scratching deep between sand and stone to get at the precious liquid. Wild horses would paw away the gravel, watched by thirsty birds perched on the Dead Finish. Eagerly they would flutter down and gulp their share, thrusting their thirsty little beaks right under the wild horses' noses. Sometimes a cloud of chirpy finches would be like flies squabbling under a horse's nose with him snorting to blow them aside.

At dawn we would be warily climbing among the rocks, often to come face to face with a startled wallaby, he wheeling around as a lad whipped up the rifle. A wallaby taken thus by surprise may react as comically as a man. In ordinary ways also. The wallaby may be reclining in the shade, lazily contented and not thinking of anything in particular. He screws up eyes and nose, nearly yawning his head off. Then brushes away a fly with an irritated flip of the paw. Dreamily he scratches head or ribs; rises on an elbow, gazes around at nothing in particular, then lies back again with half-closed eyes. Presently, he'll fall into a snooze.

Often we tried the same game on wallabies as on the larks – freezing. "Freeze" to rock or tree, to bush or earth. Rendered uncertain by such tactics, a distant, uneasy wallaby will stand bent forward to hop, staring anxiously around, his ears twitching to catch the air-currents that might carry sound. Down in the gully and across on the hill opposite he sees nothing; hears nothing suspicious. It must have been his imagination! From the hollow of a gnarled gum down below comes the hoarse, gluttonous gurgling of baby galahs, while a yellow-throated minah chatters to his friends. All, apparently, is as it should be: sun-splashed rocks, a wedge-tailed eagle in the sky, the slither of snake over pebble and stick, the listening silence of the bush.

The wallaby sits back erect; all seems well. Hesitating, he bends forward and slowly hops upward, climbing from rock to rock. Then the stalker crouches swiftly forward to freeze again as the wallaby, sitting on a rock commanding a wider view, gazes back along that harsh, shadowed gully. But sees nothing; hears nothing. His ears, the poise of his head grow less alert, he lowers his muzzle to bite at a tick then licks smooth the fur before glancing around and listening again. At last he hops leisurely away to resume his foraging, or seek a favourite shade. The stalker swiftly follows him.

But should a startled wallaby's attitude betray intense distrust, he will bound swiftly forward hopping from rock to rock past bush after bush with a thud, thud, thud, until he disappears around the hill.

There is one chance of stopping him – whistle shrilly as he bends to hop. Instantly the wallaby sits upright, a picture of startled alertness pulled up short. Ah! that whistle was not of the native bush! He calls upon sense and nerve to locate that sound. Then the stalker, if there be cover, crouches forward. If none, he waits until the wallaby bounds away, then races forward to the nearest rock. Shrilly he whistles ... The wallaby straightens; wheels around. The stalker then crouches forward, drawing closer from cover to cover. It all depends on the stalker's skill.

It is possible to stop a wallaby thus at such a distance that the animal only faintly hears the whistle. Then patiently trail him until they meet practically face to face.

To secure a euro skin was a triumph. The euro is a nuggety, fightable chap, a cross apparently between a kangaroo and a wallaby. A shaggy animal, extremely shy and unapproachable, keeping to the quietest and rockiest hills. To spy him outlined a-top of a jumble of black rocks is faintly reminiscent of a "tall" pig squatting on its haunches. For some boyish reason, we always liked them. They were defiant, game; and among themselves great fighters, the bosses of the range.

Living in the Silver City, in an atmosphere impregnated with minerals, working in the Big Mine, now tense with the great experiments, we boys on our wandering expeditions kept an eye on the rocks, picking up any likely looking stone for closer examination back at work. One day a stone caused trouble. It was a faintly greenish, yellowish looking stone. Jimmy Trembath, the boss assayer of the Fire Room was interested. I crushed it; Jimmy assayed it. It went rich in silver. One of the mill bosses happened to stroll along to the Assay Office while we were discussing the assay. Jimmy showed him the silver bead.

"Where did you get the stone?" the mill man demanded.

"Out in the hills."

"What hills?"

"Oh, out where I've been shooting."

"Where were you shooting?"

"Out in the hills."

"Look here, my boy, I want a direct answer. Where-did-you-get-that-stone?"

"Out in the hills."

He was about to go "off the handle" when Trembath broke in: "The lads of ten bring in stones from the hills when they're out shooting. They assay them here; it's good practice and training."

"But they don't bring in specimen stones of silver!"

"No. All the same, I've seen them assay a stone going a few ounces, now and then."

The mill man kept staring like a hesitating wallaby. For days he tried to find out the locality. Had he only inquired in a matey way I would have been pleased, even flattered, to have told him all about it.

Eventually, he half threatened to accuse me of stealing a sample of rich silver chlorides, imagining that in self defence I'd give the locality. Jimmy Trembath became angry at this move, and stopped it. That mill man could have accused me of stealing the whole Proprietary Mine and I would never have told where I picked up that stone. I resolved not to go shooting near that particular locality again.

But later, when a full grown man, I'd go there, peg out the area, and search for the vein or lode from which that stone must have come.

But I never returned. To-day there may be a rich silver-lead lode lying far out in those barren hills.

6 WHEN FEVER COMES

It was near Christmas. Each night for a week we'd been sitting for our exams at the Tec. I'd been feeling wretchedly ill and fighting desperately against it, working by day and sitting for exams by night. Those who passed these last exams would hold certificates as practical assayers. A great stride up for young lads. This was the last night, the very last exam. Working grimly all day, hanging on until night-time, had been torture to a feverish body and a mind that simply would not think.

Walking home after the exam that night I wondered what on earth I had written down; vainly tried to remember the questions, the garbled foolishness I had written beside each. This last exam was Theoretical Chemistry, my weak point. It was late; only occasionally the muffled footsteps of a pedestrian were heard. When turning into Zebina Street, though, two men emerged from the darkness and each side-stepped. Vaguely I realized they thought me drunk, for I was swaying along from the footpath to the gutter. Then I saw the snake, a monster black snake wriggling across the footpath to the road. I stood and pointed and shouted to every one about the snake. Then suddenly stood silent – pointing. There was *no* snake. I glanced around; no one had heard. Very frightened and quiet I hurried on up the street. Home was very near.

In a few hours I was delirious with typhoid fever.

Mother nursed me, so someone said, for a week. She refused to allow me to be taken away. I only remember that at last the ambulance did come, not the ambulance of our day, but a little canvas-covered stretcher on wheels. Mother was crying as I smiled good-bye; dad and the ambulance man pushed the little ambulance all the long, long way through the streets to the Big Hospital.

It was weeks before dim consciousness came. I did not want to die; determined I would *not* die. But feeling so forlorn, so utterly hopeless, at last I asked help of the power that has given life to everything. I remember becoming gradually conscious after long intervals in that lonely hospital ward, packed with beds so close together. I distinctly realized I was far away from all these patients, in a totally different world, a tiny, tiny life wrapped in a silence far, far away. And I'd lie motionless fighting for hours alone, grimly hanging on to a flickering thread of life.

I was given up for dead three times. It may have been a wonderful constitution, the constitution building up since a toddler of those leagues and leagues of walking and climbing hills and of boyhood days in the bush. Or it may have been definite help from something immeasurably different. I'm firmly convinced it was. The constitution of a thousand thousand elephants is as nothing without the spark lent by the Giver of Life.

A day came when they carried me from the Hopeless Ward to the Ward with a Chance. I grew rapidly better. Powerful men were dying, all through the day and night. A grim tally has typhoid taken of the Hill. From the days of the prospectors when they perished like dingoes beside hollow logs; from the times of the Great Silver Camps when they perished on the track, beside their swags, collapsed in the stinking mud of maggot-filled waterholes, or died in bough shed and shanty and camp; from the time when they fell from their dying horses even, right through past the days of Umberumberka and Apollyon Valley, Silverton and the Pinnacles, until a comparatively few years ago typhoid has scourged the Barrier.

The Ward with a Chance was life compared to the Hopeless Ward, but grim enough for all that; men kept dying, slipping away out to the Beyond. And the ward was so crowded that overworked doctors and nurses could not always keep the cases separate. Doctors and nurses and wardsmen were bloodshot eyed and haggard, dropping on their feet. In an epidemic such as this every man simply had to take his chance; they put a screen around the bed of a dying one, his last privacy. The other patients then could only hear, not see. When they woke up in the morning the screen would be gone and another patient would be delirious in the cot. We were clinging so fiercely to the gathering threads of life that we daren't for a moment let a dying man frighten us; we were fighting against the thing that had him by the throat. I was very sorry for one young married man with raging pneumonia. The screen was around him. Between gasps for breath he was trying to console his sobbing wife and mother.

The Convalescent Ward was a paradise through which Demon Hunger stalked; a convalescing typhoid patient famishes for food and food and food. He is starved so that, as the weeks go by, he will implore for food; will steal it with the cunning of a madman. His eyes will haunt the nurse, imploring the crumb from a biscuit, a crumb from anything at all!

But even a tiny crust means death. How the matron and nurses used to watch us! It was a terrible hunger. It was my first experience of nurses. My other experiences have been the same. And yet, a nurse on duty is a friend indeed.

Dad came when I was convalescing. Visiting days were paradise, with long, lonely intervals between. When the two precious hours came we felt we were alive and back in the world again.

I was surprised that mother did not come. But dad explained that she dare not come yet awhile, for fear of taking back germs to my little sisters. ... I grew stronger and stronger-and yet mother did not come. I began to get frightened ... At last they had to tell me. Mother was dead. She had caught the fever through nursing me. I tried to die too.

S.S. *Newcastle,* in Newcastle Harbour.

7. BACK TO THE BUSH

Dad took me straight to Sydney, to – the old gran's, a train ride then through three States. Unhappy days, waiting at gran's for returning strength, longing to get away – anywhere. Canny old gran saw here a grand opportunity to train up another man about the place; to pack him off to office in the morning with his tram-fare and sandwiches for lunch, and see him come back each evening on the minute for supper; then a talk, then bed. To manage his pay envelope, dam his socks, keep his collar clean; in short, to "look after him".

She had thus tamed old granddad, much to his eventual benefit. That boisterous old boy had been a bit of a rip snorter; he'd drive a four-in-hand any day rather than two, and never let the dust catch him either. But when the little old step-gran hitched up with him she put the brakes on, much to the giant's surprise. She civilized him; saved and invested his money and assured him moderate comfort in his declining years.

Had I knuckled under to gran I would years ago have been a wealthy and a crabby man. I'm bad enough as it is.

One day I turned into Sussex Street, haunted the wharves and got a job – lamp-trimmer on the old S.S. *Newcastle*. It seemed the breath of life again. I hurried back to Waverley and packed a few clothes. Gran's wrinkled old face set hard as the Sphinx. She forbade me to go; threatened to call in the police. I grinned, kissed her leathery old cheek, and hurried back to the wharf, just in time to sail.

A grand adventure – if only I'd had the strength, instead of being a pale little bundle of skin and bone. The lamp-trimmer's berth was in the tiny bunk hole of the gruff old bosun. I was bosun's mate and lamp-trimmer; the crew a very decent little crowd. The lamp-trimmer's job among other duties was to clean the ship's lamps; put up the starboard, port, and masthead lights at night; lend the bosun a hand, and take a turn on watch. It was thrilling, on black nights, swaying up there in the crow's-nest; ringing the bell at intervals with the call: "Eight bells sir! and all's well!" watching out for a disappearing and reappearing star to call: "Light on th' starb'rd bow sir!" The small ship was an intimate live thing pulsing underfoot with the masthead silhouetted as it swayed between sky and stars. One dark night I had to crawl out along the side of the ship above the waves like a monkey clinging to a swaying guttering on a windy night. The paddle-wheels were thrashing away with a mighty humming, bashing into the waves, chewing them into bubble and foam. Suddenly a big bat-like thing brushed past my face; the shock almost loosened my grip. It was the ship's black cat. What possessed her to spring straight out from the deck on to the paddle-box goodness only knows.

The food was quite good. And the fo'c'sle reckoned it must be worth while getting the fever, to develop an appetite like mine.

But the bugs! They started to eat us alive immediately we turned into bunk. But a man must sleep, for soon he must tumble out on watch again. The power of deep sleep will defy even bugs.

Back in port, I stumbled off on to the wharf, crawled home to gran's, and fell into bed. A relapse. Weeks of misery recovering from that. Gran wore an "I told you so!" expression. Sullenly I wondered if I was ever going to regain strength to go right back to the bush. But how could a penniless lad strange in a big city find a job in the bush?

Gran specked numerous chances in office boy jobs with "old established reputable firms" from whence a quiet, painstaking, obedient lad could work up to suburban respectability. By a hairbreadth I missed being cabin-boy on a German steamer bound for Hamburg. I signed on, quietly rolled my things, and hurried down to the ship. But gran beat me to it; raised Cain at the shipping office.

An Employment Agency advertised for station-hands and rouseabouts. "Only experienced hands need apply."

"You've no earthly chance," insisted the employment agent. "You are only a weedy lad, a sickly one too; the bush would kill you."

I battled for a job, any job, anything away from the city.

"I tell you lad that even if I had a job I daren't recommend you! You've absolutely no experience. Remember, these station-owners are my clients; if I sent you to a job they would know you were unfit within an hour. Then I would lose a client, the squatter would never again engage me to sign on labour for him."

I insisted. "Come back in a week," he snapped. A week later he said: "There's a job for a boy on a selection away out towards Moree. Fare paid. Sign on for six months. Pay is five shillings per week and keep. Will you take it?"

Five shillings per week! At the Assay Office I was drawing thirty shillings per week, on the S.S. *Newcastle* thirty-six shillings per week. I'd thought I'd left the five shillings per week days behind with the school trousers. But – that chance to get to the bush! Then, six months later free to go anywhere. Anywhere in the bush would be home.

I quietly packed the old portmanteau, left a note for gran, and caught the train. In early morning I stepped out of the train at a tiny siding. A big, well-dressed, important looking man, adorned by a white helmet, stood talking to the station master. The big man was my boss. A bush store and pub marked a junction of roads at this siding. Probably a small township has sprung up there since.

The boss drove me out to the selection. Past a station homestead, then out on to the great black-soil plains where, at eleven miles, the roof of the selection shimmered in the heat. The boss was not the type generally imagined as the "cocky" of the bush. Except when especially busy, all he did was to ride around his paddocks if he felt like it. But he knew his job.

The selection, except on the distant twisting creek, was treeless. The black soil gave abundant and sweet grass. The few Bathurst burrs and Darling peas and other pests I kept in check. Except at mustering-time, lamb-marking, and shearing the boss sensibly led the life of a "gentleman cocky". He was well looked after by a housekeeper, a neat, busy little woman. She cooked for us both, kept the "front house" tidy, and managed the butter, and fowlyard. A hard-working widow with a little girl, her life was "station work", her ambition to save enough to give the child a start in life. The big slab kitchen with its several rooms was at the back of the "front house". I slept in a cubby hole in the kitchen, quite satisfied. On wintry nights the wind would moan dismally in between the cracks in the slabs, bringing an icy song to face and ear.

A willing lad will soon master anything. I'd milk the cows in the dawn. After breakfast the boss in heavy, solemn voice would give his instructions for the day. Then he'd mount and ride majestically away to see how the wool was growing on his sheep's backs. I'd saddle up horse, or put it in a cart, then out to a back paddock to repair or erect fencing or stockyards, skin a dead sheep or two, or hoe Bathurst burrs or the poisonous Darling pea. Learned, too, to spot a "killer" – to tell whether a sheep was good mutton by feeling its tail.

Some among the birds of the creek, the crows especially, soon got to know the human who weekly killed a sheep!

The crows used to pass on word that I was coming; I got to know some among them individually, even though in flight. When passing overhead they'd invariably look down with beady eye and kark a most understanding kark.

Work on the selection was excellent training, quickly bringing back strength, even though it meant toiling from dawn to dark. It was lonely enough. Apart from the unapproachable boss there was no man to yarn to. The great plains under brilliant sunlight seemed awfully big. A drover's outfit passed the house once, fortunately when I was there. Otherwise, except when I rode to the township for mail, the only man I remember speaking to in all that time was a boundary-rider from Mascot station. As a gigantic, dancing blur he rode out of the heat haze one stifling day. We yarned for several hours; then he rode his way, I mine. That yarn with a matey human marked a real red-letter day.

The housekeeper was a cheery little woman; but she must have been lonely. After the evening meal we would yarn for an hour then, healthily tired, each would seek sleep preparatory to another long day. We were fond of cakes, which were not always forthcoming, as she did not care to use too many of the boss's eggs. I solved the problem when bird-nest season came on.

A merry trio of whispering conspirators, the housekeeper, the little girl, and I.

"I'll rob the poor birds' nests for eggs to make our cakes."

She laughed.

"But Jack, birds' eggs won't make cakes."

"Why not, so long as there's enough of them?"

"They're not fowls' eggs!"

"They're eggs, all the same."

"But they're so small; it would take a bucketful to make a cake!"

"No jolly fear. Hawks' eggs are large eggs, and there are plenty of them. Crows' eggs are fairly large too."

"Crows' eggs! Those filthy brutes! And they pick the eyes out of the poor lambs too. Don't you dare ever bring any crows' eggs here, Jack."

"Right-oh! All eggs go in the cakes except crows' eggs. But watch out I don't slip in some goannas' and snakes' eggs."

"Jack! That's the finish of the cakes! I'll have nothing to do with it."

"That's all right," I grinned. "The snakes aren't laying now any way."

So one morning I found it convenient to ride the creek searching for bogged sheep. The long, winding creek held the only trees on the plain, so here birds must come from many miles around to nest. Apart from the wee things that honeymooned in the rushes along the waterholes, the birds were fairly large, from magpie, owl, cockatoo, galah, hawk and crow – whose eggs the housekeeper wouldn't know from any other eggs – right to eagle-hawk eggs. Nearly every tree contained from two to a dozen nests. They were all large trees; it's a wonder I didn't break my neck. At sundown I returned with a billycan full of eggs. The housekeeper was all eyes.

"What funny eggs, Jack! Look at the pretty colours. But the size! Why, it will take me an hour to mix these little things into a cake. Are you sure they're fresh?"

"As daisies," I replied confidently.

"I'm not so sure. Whatever will the cake be like. I'd better hide them away until to-night."

That night, we three whispered over the eggs. There were green ones and white ones, pale blue ones and red-speckled and brown-splotched ones, product of parrot and magpie and hawk and eagle-hawk, owl and galah, crow and cockatoo. The conspirators were agog to know whether the eggs would make cake.

They did. We enjoyed a feast the next night. Thereafter the housekeeper had no doubts, whenever opportunity occurred her eyes and smile suggested I slip away to the creek for eggs.

It was good until I'd worked out the creek for miles. We got many eggs, made and ate quite a number of cakes. In fear and trembling the housekeeper one evening gave the boss some of the cake to finish off for his supper. He complimented her on her skill as cake-maker.

Lightning Ridge, now.

8. I RUN AWAY

Five months went by; then trouble came. The cows were not giving enough milk; hence shortage of cream and butter. The boss complained; the housekeeper grew anxious. I explained to the boss that he had given orders that no fresh cows were to be broken in. Irritably he replied that the old cows were as fresh as ever and that it was my fault they were not giving enough milk. I replied that I was not a bull. Angrily he answered; I shrugged in reply.

He then accused me of not milking the cows right out, of letting the calves take the milk so that I could dodge milking them.

Next morning I determined he should have his quantity of milk so put the half-empty buckets under the tank and turned the tap on. The housekeeper was very relieved when she saw her big milk dishes full again to overflowing. At breakfast she glanced at me as if suspicious. I really had been shirking the milking, allowing the calves to suck their mothers. Next morning there was hardly any cream on the milk in the big dishes; the boss would have to go without his butter.

"What can be the matter, Jack?" asked the troubled housekeeper. "The grass is not going off."

"No. There's plenty of grass."

"But there's no cream!"

"No, and there won't be until he milks fresh cows."

And so it carried on until the fourth morning. I was standing by the tank, with the tap full on into a bucket of milk.

"I thought so!" roared the boss. His look threatened to pull the place down.

"What do you mean by this?" he roared.

"Nothing. Only giving you your milk."

"Cut out that cheek, you infernal brat, or I'll tan your hide so that your own mother won't know you."

"Try it."

He stepped forward, then hesitated. It would be a long time before he would get another lad to do a man's work for five shillings a week. Had he made a grab I would have slipped around the tank to where I'd hidden the axe. He could have stood back and used his stockwhip but I would have run in and chopped him.

"Get to your work!" he roared.

"I've finished. Give me my time."

"Oh, that's how the wind blows, is it? Well, let me tell you you're not finished. You've signed on and you stay here until your last hour is up."

"I'm going."

"You're not. You've run away from home. *I* know. And if you run away from here I'll ride straight into Narrabri and have the police after you. I'll have you back in double quick time."

White with temper he stamped inside to a spoilt breakfast.

I was surprised. Gran or someone must have got in touch with him. That explained why he had allowed me to ride occasionally into the township for the mail. He must have thought I was an outcast; daren't leave the place.

That night was pitch dark, though far away up there twinkled a million stars. I did not wait until his light went out. No jolly fear! but crept out the back while the light, but above all an occasional grunt, betrayed he was in his room. Straight out past the woodheap, then aside well into the lucerne paddock, then back towards the track that led to the township. The railway line would be a guide, going north farther into the bush.

The portmanteau was a tragedy both to carry on the shoulder, and to own. To carry a port in the bush was ridiculous, apart from being a big physical strain. But I was determined he would never see it; I would leave and hide it among the first trees, miles away.

It was desperately lonely under the stars on the big plain, with ears straining to catch sound of a horse coming behind. But there was a feeling of greatness too; the great plain, the great sky, the vast night and a midget lost it seemed. Once past the station and he would never catch me – nor the police either. Surely the police would never come into it. And yet, I had signed on for six months. What really was the law? Whether his threat of the police was sheer bluff or not I neither knew nor cared.

In the small hours there appeared the far-flung shadows of the station homestead. The rambling buildings lined one side of the road, the shearing-shed and men's quarters the other. Beyond was a creek which the road crossed. It was a case of following the road between homestead and men's quarters, for to work far away around meant striking a deep creek with no crossing.

The dogs had me scared; they were snarling now. If they came rushing out, they would bail me up, if nothing worse. The station people would then come out and I would be caught.

Bending low and resting to the earth every few yards, I crept down the road. Never did hunter stalk a kangaroo with as noiseless care as I crept down that road, and never had shadows been so welcome. When passing between outbuildings and homestead the dogs, much closer, were snarling among themselves away to right and left. But whatever worried them, it was not I. At last I came to the big white gate. Through that lay safety. Noiselessly, I opened the gate, lifted the portmanteau through, then hurried silently down towards the creek crossing.

I can still remember the breathless feeling on straightening up. Danger was behind, the world lay ahead. Soon appeared dark masses of timber. In amongst the shadows I knelt down in a patch of starlight; tipped the contents of the port into the blanket; rolled it into a swag with a towel as strap; then hid the port under a clump of branches. Picking up the swag I hurried back to the road, and before daylight was on the railway line, heading north. It was a glorious dawn; the sun all rosy gold, birds singing, two steel rails running straight ahead. I laughed to the bush I loved.

In late afternoon the shadows lengthened, coolness came, the birds livened up, getting the best out of the last of the sunlight. Evening again brought the loneliness of the bush. An owl swished by, his eyes golden balls as he glared down. In the darkness a flower of flame appeared bobbing among the tree-trunks away to the left. A camp-fire. Hopeful and ravenously hungry I turned towards it, noisily treading on the twigs. The flame grew, the white of a tent showed up, then the half finished framework of a new house. A fairly tall, strongly built man sat smoking by the fire. The flames lit up his expressionless face.

"Evening," he murmured.
"Good evening."
"Travelling?"
"Yes."
"Care for a drink of tea?"
"Thanks."
"Billy's on. There's a pannikin; help yourself."
I slung down the swag with a smile. Presently, as he noticed how the

tea was appreciated, he added: "There's damper in the camp-oven, and some shoulder of mutton in the safe. Help yourself." He puffed his pipe. I ate. Then: "There's plenty – and plenty more where it came from. Go ahead."

I accepted with the famished hunger of youth.

"Ah, that was good. I'll put the billy on and wash up."

"Good-oh."

I washed up. Then he said: "Have a smoke."

"Thanks." I lit up.

"Come far?" he asked.

"About thirty-five miles."

"Looking for work?"

"Yes."

"Ever done any bush carpentering?"

"No."

"Doesn't matter, I see by your hands you're used to bush work. The boss wants a young fellow to give us a hand with this cocky's house we're building. Pay would be thirty bob a week, and keep."

"I wish I could get the job."

"I'll speak to the boss in the morning."

Thus I got the job of labourer to a bush carpenter. The boss was one of those small, efficient, quiet men. While not appearing to do so he watched closely a new hand at work. If the new hand did his work the boss said nothing. If the new hand was a loafer, the boss quietly gave him his time.

We worked steadily all through the days and it was surprising how swiftly that house took shape. All bush timber, too. Evenings were filled with pleasant contentment, after the solid day's toil.

I camped with Jim. I've met similar Jims since; quiet, steady, dependable Jims. This Jim cherished a secret. And later by the evening fire partly confided in me. He was in love. But he had done something far away (he would never quite tell what) and was in disgrace. He was never to get his girl, but he often thought about her; he had been thinking about her that night by the fire when I strolled up. It did him good to meet a sympathetic listener. I have realized the value of a sympathetic listener, now that I have turned into a bit of a chatterer. A young fellow who is a sympathetic listener cannot help making friends in early life.

It was with pride, tinged with regret, that in two months time we saw that neat little home finished. The boss and Jim were breaking camp preparatory to driving into Moree to a new contract in town where a lad was awaiting them, convalescing at the hospital. It was his job I had walked into.

But the boss, on the quiet, had got me a job. He had recommended me as a steady, hard-working lad to a "big cocky" some twenty miles out. On the morning when the carpenters were to go the boss paid me two months' money at thirty shillings a week, the largest sum I had yet received in one lump in my life. An hour later the new boss arrived with a spare horse – a young one. Shaking hands with Jim and the boss, I turned to warily strap the swag on the saddle upon the young horse. The new boss was a sinewy little man with a grey goatee, and twinkling eyes that saw everything ...

But the young horse could not throw me.

9. ON THE TRACK TO THE RIDGE

The new job was sheer pleasure, riding about the place breaking in horses. The boss lent a hand – just as well.

The selection was almost a small station. The boss's daughters were accomplished horsewomen – to his delight. Nice girls. They had helped him when he was struggling; but now they were more given to riding into Moree and doing "the grand" as the boss phrased it. After having toiled hard, why shouldn't they enjoy themselves. A rouseabout and several station-hands were employed.

The boss's joy was his horses. He bred them. Classy beasts. On most mornings a wink from the boss would mean to ride away to the back paddocks. He would meet me by a circuitous route. But the family knew what he was up to: "Away to the back paddocks again to break in horses with that young Jack!"

The family wished he would give up horses and put the money and time and grass into sheep. It paid better. Besides, he was too old now to be bothering with horses, he'd meet with an accident any day.

So the old-timer with the twinkly eyes used to wireless his orders with a wink. We'd meet behind the wool-shed with tucker in our saddle-bags. Then, with pipes alight, seek the bush at the amble and run a young mob into the stock-yard.

Breezy young colts; clean-skinned bays and chestnuts and greys bucking with life. We'd rope one wild-eyed, quivering young thing and handle him. The boss had wonderful hands on a horse, the endearment in his voice told how he loved the young things; he could do almost anything with them. When we had the colt somewhat reassured we'd gently coax on the bridle with the colt leaning back trembling while we slipped the saddle on and girthed it tight. I'd grip reins and mane, slip one foot in the stirrup while the boss held its head talking to it, then the instant I was in the saddle, he would let go and the colt would begin frantically pig-rooting around the yard.

To a very occasional buster, the boss was training me as he trained the horses. He picked out those he expected would buck the least, and these were ridden first. Each young colt and filly broken in and ridden completed a new lesson, brought experience and confidence. Thus I learned to really ride.

Having mastered a colt in the yard we'd ride away bush, the boss sitting his steady old horse, I on the young thing itching to root every hundred yards. And the boss was proud as a king; proud of his young colt, proud of me. He was a great old scout.

One evening several travelling shearers arrived. They'd been putting in time at a new opal-field called Lightning Ridge, away west. At the men's hut they yarned of "nobbies", and "fire", and "harlequin", and rich parcels and dazzling money awaiting the lucky "bottomers" of "holes".

"Some blokes strike it lucky first go off," they explained; "others might work twelve months and not strike a colour."

"Where is Lightning Ridge?" I asked.

"Only about a hundred mile west as the crow flies. From here a man would make for the Gwydir. Follow it down a bit, towards Collarenebri. The Ridge is about fifty mile nearly sou'west of there."

"I've a good mind to give it a go."

"Why not? You only need a few pounds for tucker money. Sink a few shafts. If they're duffers, pack up your swag and go; there's plenty of work in the back country. If you're lucky, you'll make more money in a week than you'll make in three years on a station."

That chance meeting stirred the mining fever quiescent since leaving Broken Hill. I determined to save a few more pounds and "give the Ridge a go".

Only a week later the boss started me off, the last thing in the world he intended to do. I was riding the old cock-eyed grey night horse, the boss ambling along with his pipe alight. We had sneaked away to ride the paddocks and break in a fiery young chestnut.

"You're a nice one," he puffed in injured tones.

"Why, boss?"

"You never told me you ran away!"

I stared.

"I hear that old — way back at — is raising Cain," he went on after a while. "Says you ran away from his place without finishing your time. Says now he knows where you are he'll have the police on to you and bring you back."

"I've finished with —," I answered shortly.

"It's all right, Jack," he winked, "— has got a dashed cheek, I won't let him interfere with you anyway."

I rode on silently. My opinion of grown men had collapsed. Both these men had tried to frighten me with the bogy of the police; the one because he had been paying a lad a schoolboy's money to do the work of a man; the old boss here because he liked me and wanted to keep me.

"I want my time," I said.

"What?"

"I'm leaving."

"What for? What's struck you!"

"I want my time at the end of the week," I replied. "I'm going."

I did. And another walk in life began. On "the track to Merrywinebone". But I didn't care much where the track led. The world was young and brave, the bush was beautiful. If a man could pick up a job or two and increase the tiny bankroll, then, so long as he was moving towards the Ridge everything would be quite all right.

I boiled the billy beside a quiet stream, eyed suspiciously by a married magpie in a nearby tree. A crowd of Noisy Jacks soon gathered around and started a loud-voiced argument as to this human feeding by their creek. A damper crust divided among them caused hilarious excitement. The days had gone when I would hurt a bird. I loved them.

Trudging along, loneliness came with sundown. The howl of a dingo, the croak of a mopoke, the stars and camp-fire and pipe were company for the night. Early next afternoon a horseman emerged from a clump of belah:

"Good day."

"Good day."

"Makin' anywheres in particular?"

"No."

"Lookin' for work?"

"Yes."

He half turned in the saddle and stretched a brown arm towards the south-west.

"Straight in through them trees, about six mile, there's a gang ringbarking. A coupla days ago the boss mentioned he could do with another man or two."

"Thanks. So long."

With a farewell grin I turned into the timber and presently located the gang taking the timber in a face, each man busy about a hundred yards apart working with a light drop-cut while walking around the tree, each man toiling methodically. They get over a lot of country that way.

"I'll give you a start in the mornin'," grunted the boss. "But we're a slick gang here."

"Thanks. I'll keep up with them."

"H'm," he grunted and was already ringing another tree.

A happy camp, twenty-three of us. The boss, the Babbling Brook, and the gang. Pleasant but destructive work. Killing hundreds of trees every day, and laying the foundation of a nation-wide erosion. Every day, throughout settled Australia, thousands of axes kept rising and falling;

thousands of trees were dying to every tick of the clock. If a machine could be invented which could recall those axe strokes, then release them in one concentrated blow, what a terrible roar of destruction would result!

The white-whiskered old Babbling Brook used to make scrumptious brownies – that big, sweet, brown cake fresh from the camp-oven. We were craftily diplomatic in our dealings with the Babbling Brook. All bush cooks are touchy; they know they've got you where you feel it most. Our old cook was a good old sort though, so long as we didn't leave anything lying about in his spotlessly swept bough-shed kitchen, and sat down to meals on the very tick, and played no jokes on him. I've seen a bush cook go raving mad simply because some joker mixed Epsom salts with the baking-powder.

The quietest man in camp was Big Bill, one of the best grass fighters in the back country. He and another bare-knuckle man had fought themselves to a standstill after three solid hours of slogging, staggered into the shanty for a feed and a rest, then come out to it again and fought another hour in the moonlight, until Bill knocked his man.

With drink in him he was a maniac, but in camp he was a demon worker, pleasant and quiet. Old Stretch'em was the camp character, a little old bearded chap who listened while puffing at his pipe. At just the right time in a soft, drawly voice he'd break into a yarn and gravely carry it to some preposterous conclusion. After the evening meal we all gathered around the fire, pipes came out, yarning commenced. One evening the boys were discussing a fight they had watched that morning between a goanna and a snake, which had ended by the snake swallowing the goanna.

"When you come to think of it," mused Snowy, "everything seems to live by swallowing something else. Birds swallow seeds, sheep eat grass, we eat sheep. It's a rummy go. Perhaps that goanna had swallowed a bird, then the snake swallowed him."

"It works out that way," drawled old Stretch'em; "it's a fact a' nature." Removing his pipe he carefully spat in the fire; "I remember one summer I was camped in the Barwon. A fly settled on the blanket an' no sooner began lickin' a speck o' sugar than a brown spider hopped on him an' scoffed him just as a big hornet pounced on *him*. Jest then one of them little baskin' lizards sprung on the hornet, an' was lickin' his chops when a bird dived on him, only to be pounced on by a hawk that was nipped in the bud be a goanna that swallered 'em all.

"Jest then a snake sizes up the goanna, an' they started a ding-dong go. The snake was just swallerin' the goanna like we seen that snake swaller the goanna this mornin', when a whopper snake comes slitherin' out o' the grass. He gets the other snake's tail down his neck an' begins swallerin' him.

"It was a great go, but it was all up with the little snake wot had his jaws nearly busted with the goanna that had the hawk that had the bird that had the lizard that had the hornet that had the spider that had the fly that had the sugar. At last the big snake swallered 'em all an' lay there helpless an' bloated like a balloon."

"And what swallowed the big snake?" asked Snowy sarcastically.

"A wild pig," answered Stretch'em. "He came with a rush an' got inter that snake. But somethin' must 'ave tickled his insides, because he let out a yell an' jumped into the river fair into the mouth of the biggest Murray cod you ever seen. That fish 'ad jaws on 'im that could just about 'ave swallered a battleship, but the pig with all he had inside him was nearly too big a mouthful. The last I see of 'em was the fish floatin' down stream with the pig's tail wigglin' out o' his mouth."

"And I suppose," said Snowy deliberately, "that when they floated down out to sea a shark swallowed the fish and a whale swallowed the shark!"

"Somethin' like that," nodded old Stretch'em gravely. "Once I noo a bloke called Jonah ..."

Opal cutting at Lightning Ridge, 1905.

10. THE LOONY BOUNDARY-RIDER

When the ringbarking job cut out there was a cheery reining in of horses, rolling up tents, packing the camp-gear into the buckboard and cart. The gang were all going back to Moree to spell-oh before starting out on another contract. They pressed me to join in with them; we could always get a job around the Moree district; – I'd soon "get known".

That sense of security was tempting. Bush workers often make a country town their base. They "get known". Men and gangs work the year round that way. But I was growing more confident with each job; Australia was a big place and it would be wonderful to roam all over it. I waved to buckboard and horsemen as they vanished through the timber then trudged off in the opposite direction. With a swag loaded with tucker including a huge brownie the Babbling Brook had especially cooked.

I was following a bridle-track. A station-hand had volunteered directions. "To go around by road," he drawled, "would be about a hundred miles further. So long as you can travel through bush you'll be all right. Follow that track out past the Six Mile Tank. It's old Tom the boundary-rider's track; his hut is away out on the northern boundary. Tom takes his stores out once a month by packhorse, an' that's the track you follow. It's a long day's walk though, you won't make it by sundown unless you step out. You've got to go through a belah scrub too, and that's a nasty place to camp in. If you make Tom's hut by sundown you can camp with him; he ought to welcome company anyway. He's a queer old bird, been camped for years on his own but he's all right if he takes a liking to you. And whistle before you get to the hut – he's got a savage dog."

"H'm. I suppose I whistle so as not to come on him by surprise."

"You've guessed it," grinned the station-hand. "Some of these lonely old blokes are a bit queer; you never know what they might be doing if you come on them suddenly, and they don't like it."

"And if the dog comes at me it means that the old chap doesn't want visitors."

"That's about it. Get a whistle up before you come out of the belah, then stroll towards the hut. He'll call the dog to heel."

"H'm. And how do I travel from there?"

"Go bush directly north about five or six mile, the country rises a bit then. At five mile you ought to see a couple of low hills away in the distance, a bit to the left. Take your bearings and make for them hills. A track runs in front of them, which leads into the Collarenebri road. Follow the track west and you'll hit the Collarenebri road."

"Good-oh. Thanks. So long."

"So long. Good luck."

By mid afternoon the swag had grown jolly heavy. I was a bit uneasy too, for the track had deviated at an angle veering farther and farther away. It was only a horse-pad; but there was only this one track, so the station-hand had said. A mile farther on and the track spread out, faded, vanished. I'd been a careless fool. This was only a feeding-pad made by the boundary-rider's horses. From the hut they would walk back along the track, then branch out here on this pad they had made themselves and thus wander towards some favourite feeding-ground.

Hurrying back I rejoined the track and pressed on, having gone three miles out of the way. Soon appeared the dark mass of the belah scrub with the track leading into its shadows. Before sundown it was about dark in the maze, the track difficult to follow as it wound among the gloomy trees, the ground a brown cushion from the countless needle-like leaves. Sundown came bringing that shivery, whispery sighing inseparable from the belah. It's a lonely place, a belah scrub by night. I hurried on, peering down at the faint impress of the track so covered by freshly fallen leaves in places that it temporarily vanished. Soon the forest became black shadows eerily pierced by filtering moonlight.

I walked now only by groping, "feeling" the track underfoot. Tree-trunks loomed threateningly like gnarled figures stretching out to push me back. Above, a faint breeze shivered the branches with their countless moaning leaves. I listened, hoping for the echo of an axe, bark of dog, any distant murmur of life. But nothing – only that sighing, rising from somewhere above, coming from somewhere around, sighing away among dim trunks, vanishing into darkness. I seriously considered camping there before I became lost. But such a lonely place to camp, no water to boil the billy either. The boundary-rider's hut could not be far away.

The sharp yelp of a fox startled me nearly out of my skin. Breathing relief I pushed slowly on remembering vivid stories of the Red Terror, a notorious dingo. This cunning wild dog had shown almost devilish intelligence in the ways of humans; the cleverest trappers had so far failed to kill him. He could not harm me of course. And yet ... this lonesome place and the uncertainty pictured his green eyes shadowing my creeping footsteps. One snap of his fangs could rip open a sheep. Might he not try it on a human?

I'd just decided to light a fire and camp when a piercing howl raised the hair on my head, followed by a mad burst of laughter that left me crouching behind a tree. The peal of laughter ended in a wild shriek. Someone was being murdered in the belah scrub!

There followed long minutes of terror. Then, a hoarse shout of command, a rollicking old English song. And silence. I crouched down on the swag, breathing painfully. The mad boundary-rider! He must be loony. His hut must be just ahead, out beyond the belah. At last I picked up the swag and cautiously groped ahead to get out of this awful belah and into the sweet, open bush. I'd sneak past the boundary-rider's hut, put a lot of space between it and me, then camp at the first water available.

Presently, a filtering of strong moonlight just ahead told of open country. A twinkle of blessed stars and a moment later I stood among the trees, staring across at a clear space ahead with the dark hut bathed under a glorious night sky. Faint light shone through a chink in the hut. I grasped a stout stick in case that dog should come sneaking out. From the hut came a resounding thump, a hoarse shout of command:

"Up boys an' at 'em. Foller me! An' fill their breach with yer English dead!" (Henry V leading his troops to the breach at the siege of Harfleur!) "Foller me, dear friends!" roared the voice. "Once more to the breach! Stiffen the sinews! Work up the blood! Howl like the tiger an' foller me! Get into 'em!"

Still no sign of the dog. He must be the audience. I crept towards the hut, then stealthily bent eye to a chink. The hut was lit by a slush-lamp, helped by the steady glow of the fire. Lying by the fireplace the sombre cattle dog stared unblinking at his master who stood in martial attitude with firelight glowing red on his face and heavy red beard. His eyes stuck out like onions as with waving arms he roared on his men. Old Tom the boundary-rider was Henry V at this very moment.

Flinging high his sword arm he pranced and suddenly yelled: "Charge, me lads, charge! Fill the breach with yer English dead! Up boys an' at 'em! Inter 'em with yer halberds! Slice 'em in the guts!"

It was a jolly good show, it held me delighted until he'd blown himself out. He came back to earth with a grunt and a grin, he puffed and ha ha-ed and clapped a hearty encore; then grinning hugely walked to the fireplace and sat down on a box. He stroked the dog's head.

"How's that, old dawg?" he grinned amiably. "Not a bad performance eh! That Oscar Asche bloke can't hold a candle to what we can do, can he? Ah well, we does have our fun. And now wot if we boils the billy an' has a taste o' brownie before turnin' in?"

I slipped away, tempted to return when the billy was boiled. But the suspicious boundary-rider would guess I'd been a Peeping Tom, and he'd be so mad he might pick up a gun in earnest. With an easier mind I took bearings north from the hut to a star and walked quickly into the open bush, dog tired. At half a mile, fading moonlight gleamed upon a gilgai hole. Thankfully I lit the fire and, smiling to its cheery company, boiled the billy ... and fell sound asleep.

At dawn, a gossipy crowd of Happy Families clamoured for breakfast. They scolded and clucked and chattered when I rolled over to sleep again. Then an old crow came and karked from a tree overhead and the family scolded him as well.

I started off happily, a rosy sun arising over the treetops just beyond my right shoulder. At about five miles the hazy crowns of the two low hills appeared above a sea of treetops. As the country dropped again those hills would soon be hidden. It meant a walk of a good many miles straight through bush, sighting from tree to tree.

Hudson Brothers poison cart.

11. THE POISON CARTS

Just at sundown the dusty track appeared ahead. It's a little thrill to walk miles through trackless bush and yet come out exactly where you wish to. Soon afterwards among the evening shadows appeared the twinkle of a fire. Another knight of the road. Cheery company. Camping alone in the bush must have been wretched for this little chatterbox, unless he talked to the trees and possums and owls. He kept on a lively conversation far into the night, and piped up again with the birds at dawn. As we walked on towards Collarenebri he chattered the constant chatter that gets on a man's nerves a bit.

This mate of a day was enthusiastic on dingo- and fox-hunting; he painted glowing tales of the interesting life we could lead, the money we could make. He had something wrong with one eye; to his everlasting regret he could not shoot straight. He admitted too he needed "a guide"; alone he could not keep to any one objective for long. He'd develop a fresh enthusiasm for something else and away he'd fly at a tangent. He wasn't a "sticker"; he wanted someone to take him in hand. Then he could do things.

He proved a wonderful mimic of animal, reptilian and bird life. As we walked along he whistled bird call after call and again and again birds from the roadside answered him. We enjoyed a smoke-oh tinder a burnt-out old tree where he called up crows until a dozen had settled on the branches overhead. Then he carried on an intimate, gossipy conversation with them which they answered in puzzled bewilderment, peering down, peering at one another as they karked: "Where *is* he?" and "Just what *is* this stranger!" Flopping from branch to branch they searched the trees around and the earth and sky to locate this intriguing crow.

While trudging along occasionally a green parrot would plane overhead. My friend would screech and the parrot would wheel to settle in a tree and screech reply.

"When I'm camped alone," he chattered, "I can bring 'em up all around me; all sorts of feathery beasties. There's hardly a bird in the bush I can't bring up. You'd imagine there wasn't a bird within miles in some places I've camped, but within an hour I've had an aviary around me. And what's more they understand a lot I say to them, and I understand a lot they say to me, and to one another."

"The gift must have been born in you."

"It was, and I've developed it since a kid. I can tell you things you wouldn't believe. I've called birds around me, and frogs and snakes and dingoes. And held 'em there too an' watched 'em eyein' one another so fearful none of 'em dared move. You needn't believe me. I'll show you one sundown, or just at dawn if we go mates together."

That evening from his swag he unpacked several clever little whistle and voice gadgets made from cigarette-tin lids. To perfection he imitated the yelp of a fox, then the howl of a dingo. He could imitate the dingo in seeking a mate, calling for the chase, in defiance, in cussedness, in boredom; could imitate an aged or old dog of either sex, or any expression in its howl, and describe the meaning expressed. I've since met an occasional man as good; waited by night while they lured dingoes right up to our hiding-place.

Only one in about every five thousand bushmen can lure animals to perfection. But this man seemed able to imitate almost everything that walked, crawled, flew, or hopped in the bush.

"Men's voices aren't made to make the sounds birds and animals make," he explained, "and that's the only reason we can't understand them. We can't talk their language, that's all, and their ears aren't tuned to understand ours. So we reckon they're only birds and animals with no sense, and they reckon we're only dumb beef. See these little playtoys? Well, they're made of cigarette-tin lids, with a tiny hollow reed inside; and inside that I've stretched little strips of tanned frog's skin for vocal cords. When I blow in certain ways on these little vibratory things they make noises birds and things can understand, noises we can't make with our mouths."

"You must have put a lot of study into it."

"None at all; I'm not educated. I just love birds and wild things, that's all."

"Yet you don't mind killing foxes and dingoes?"

"Not at all; they kill thousands of birds every year. Besides, they're bigger things and crueller things and they're chock-full of cunning. It's a real thrill to pit your wits against a clever dingo. Wait till you come to try it. And there's a fortune in it," he emphasized enthusiastically. "We'll get every rogue fox and dingo in every district that the squatters an' Pest Boards have set a price on. We'll just seek 'em out in their own wilds one by one. I'll call 'em up; you shoot them. Soon we'll own good horses and wagonette and camp, and can travel from district to district. A first-class turn-out, easy earned money, an independent livelihood, interesting work. What a life! Are you on?"

It was an almost overpowering temptation but we would not have lasted. He was born to chatter; he must chatter to live. Sooner or later that chattering would spoil our mateship. Already I knew the secret of bush mateship. If you have a real mate, you are contented. You may be two entirely different kinds of men. But so long as you can like and trust one another, can give and take, you are mates – for a lifetime if fate decrees it. But in great part you must give and take, must understand one another's moods, and set your mood in sympathy – silent sympathy if necessary – with the other man. It means only trouble otherwise. When in the bush, you must choose your mate well. If you cannot get on together, separate. Once you have found a good mate you will not want to separate.

We said "So long", at Collarenebri. He was bitterly disappointed. But it was only the mood of the moment. When I came out from the store and started down the Barwon road he was squatting over on the pub veranda chattering like a parrot to several stockmen from Dunebril station. We waved a cheery farewell.

I'd taken the Barwon road instead of the track to the Ridge because of the storekeeper's advice that there was more chance of work this way. A few more bank-notes added to that precious little bank before facing the Ridge, would be welcome. I was nearing the district now. Here was talk and rumours of the Ridge, of men who had struck it lucky, of fire stones and rich parcels, and hard luck stories and rumour stories of big buyers from overseas. In other districts the Ridge was but a name, or was not known at all.

On one side of the road were the grand old gums that followed the winding course of the slow old river. On the other was station bushland, feeding sheep visible now and then. This was a well-marked buggy road, the road from Walgett to Collarenebri. Sundowners road too; for the real sundowner followed the river down for hundreds of miles camping along its banks, working a month at a station here, a month there, then walking on again not to accept work until his funds had run out. Numbers of these nomads "run the river down" from year end to end, and rarely worked except for a week now and then for "necessaries" – tobacco money, matches, needle and cotton.

In those days it was the custom for each station to issue a ration to traveller and sundowner; a pannikin or two of flour, meat, a little tea and sugar, sufficient to carry him on to the next station. Thus the station besides being hospitable to travellers helped their wandering labour supply. There were no motor-cars those days, and labour had to travel long, lonely distances either by horse or foot. The travelling bush worker was not the sundowner.

Trudging down one bank of the Barwon the sundowners would mooch on to Brewarrina where the Barwon becomes the Darling. From

Bourke they trudged south-west right across New South Wales through Wilcannia, then Menindie, thence on to Wentworth on the Victorian border where the Darling empties into the Murray. Then they would wend their way back along the opposite bank, past Collarenebri right up to Mungindi on the Queensland border. One year for the down trip, one year for the up; crossing the State from the Queensland border to the Victorian. Living their own lives in their own way, happy men in a life apart.

One sundown I strolled in through a station gate, and up to the homestead visible from the road. The station storekeeper was issuing rations to two old sundowners, so I strolled across.

"Looking for a job?" inquired the storekeeper.

"Yes."

"Well, better see the boss up at the office. He's there now. If you don't get put on, come back for your rations."

The boss was a worker. Quick, shrewd eyes, a lined face, brusque yet kindly manner, brown beard, a sinewy frame that shouted its endurance to toil, toil, toil.

"Ever drive a poison cart?" he asked.

"No."

"Well, you'll soon learn. Used to bush work?"

"Yes. Fencing, breaking-in, ringbarking, mustering, burr-cutting, crutching."

"Right. Ask the storekeeper for rations; tell him you're going to work with Scotty. Then you'd better report to Scotty up at the shearing-shed. Tell him you're to take the spare poison cart and work the near paddock with him to-morrow morning."

"Right."

"Wages one pound and keep."

"Right."

And thus I was driving a poison cart on Woorawadian station. Scotty was a cheery soul, tall and comparatively energetic, a good looking laddie of twenty-three. With exceptionally bright eyes, his lips were ever ready to break into laugh and joke. My liking was tinged with regret, for that crimson flush on Scotty's cheeks was not the complexion of Scotland. His chest was weak.

Poison carts are of iron, just a light iron framework on two iron wheels. The poison bait is in a drum, to which an inside movement is imparted by a chain-belt connected to the wheels. The driver's seat is an iron stake supporting a small iron seat. There's a lever by the seat which drops a pointed scoop into the earth. As the cart goes along, this scoop digs a small furrow in the earth. Into this furrow from the spout of the drum drop little pellets of poisoned pollard and bran. Trailing behind the cart is a chain which as it is dragged along returns the loosened earth into the furrow and covers it up. Rabbits are naturally inquisitive of freshly dug earth. Hopping along through the bush, they sooner or later come to the newly turned earth. They dig; they smell the pollard; they eat.

The cart is springless, and being of iron can be driven through bush, over logs or small stumps if unavoidable, over antbeds, fallen branches, and any inequality of the ground, can even bump into trees, and nothing will break. When travelling through heavily timbered country the clatter and clang of our two carts used to startle the crows and 'roos for miles around.

I felt very sorry for the rabbits (phosphorus is a devilish death) but sorrier still for the pathetic bodies of birds lying beside an old poison trail.

Lamb Marking 1917, by Charles Kerry.

12. RIDING THE BOUNDARY FENCES

"This is devilish stuff!" I remarked one morning to Scotty.

We were mixing the poison mash of pollard and bran with molasses, in an iron tub. Scotty reached for the phosphorus jar. The big white sticks were contained in liquid, otherwise on exposure to the open air they would burst into flame. Scotty quickly and carefully mixed a couple of sticks with the mash which steamed; we coughed from the fumes.

"Fancy eating that stuff. It burns their stomachs alive."

"Aye, Jack, it does. An' when their wee stomachs catch alight they hie away for water."

"And the water makes it worse. It's a cruel job."

"Aye, Jack. But yon rabbits must be kept doon."

"I suppose so. But it's destroying birds too. The birds do no harm."

"There's nay many of they get poisoned, Jack. Yon chain hides the baits wi' airth again. I only see an odd dead birdie now an' then."

"Yes. But how many are there that take a bait then fly to water and die, or else die on their wings before they reach water."

"Ah weel, Jack, it's our livelihood ye ken."

He coughed violently and gasped. "Phew! Help me tip this stuff into yon drums before yon fumes burn the lungs out of us."

Man is a cruel wretch. He boasts about his civilization, and the first thing he turns his discoveries against is man. He finds that he can poison or burn or gas animal pests. Then he turns the poison against his fellow man. In the war years then coming I was to see this terrible phosphorus turned on man. Those deadly chlorine fumes in the Assay Office ultimately to be turned on man! We have no excuse.

Scotty soon allowed me to mix the phosphorus into the bran.

It was delightful driving through the bush; the birds sang; everything seemed carefree ... There were a few more dead birds lying beside a poison trail. Next morning Scotty went blithely singing, driving the old poison cart among the trees, innocent of the fact that there was no poison in the baits.

A week later Scotty with bridle over arm was passing by as I tipped the mash into the last drum. He hesitated, sniffed:

"Are ye sure ye're putting in enough poison, Jack? – two full sticks? I canna smell the beastly tack." He lifted the lid of the drum and distantly sniffed. "By golly, Jack, ye've forgotten the phosphorus," he exclaimed.

I just looked at him.

"Ye didna put it in, Jack."

"No, Scotty. I hate the thought of that stuff burning up the stomachs of the birds."

"But, laddie, we've got to put it in. We're paid to do it."

"Yes, I suppose so."

"The boss would sack us if he knew."

"He'd sack me. You've got nothing to do with it."

"He wouldna see it that way, we'd both have to go."

"Oh, well, there's plenty of work everywhere."

"Yes, Jack, for a young Australian like you who kens the bush. But me – well, I've got to take care of myself."

Thereafter, Scotty mixed the poison.

Some nights we'd fish down at the river. The big holes were the homes of Murray cods – mighty fish. Many's the fish-yarn along the Barwon. At every bend was hung up for travellers to admire a huge head of some river giant.

Generally, away around the bend a sundowner would be camping. He would fish until his flour and tea and sugar were exhausted. Then roll his swag, light the pipe, pick up the billy and casually "push on".

An old sundowner strolled into the station and condescended to accept a job at building a dray, the only work he "followed". From an old tree, weather seasoned, he cut and fashioned with station tools a complete dray that would have been a credit to any city wheelwright. A very particular old chap, with neatly combed beard, his swag was more self-contained than a millionaire's palace. All else that he wanted he carried in his mind.

He liked me because I admired his work; he merely tolerated the other station-hands. Gravely we yarned of evenings. He was a deep thinker; he should have taken a more active part in life.

"Why the futility?" he asked with a shrug. "The destiny of men and nations is written in the stars. No matter what any single individual may accomplish it is foreordained to be but a drop in an ocean. The greatest work will become absorbed, forgotten. The space of a thousand years is not a breath in time. Cities, whole civilizations have vanished, let alone the record of any life's work among their forgotten millions. Progress always has gone and always will go inexorably on, independent of what a whole world of men do for or against it. And yet, although the individual utterly vanishes amongst the countless billions that have been and are and will be, he is given a whole world to himself – his mind. What he does with his own world is left to him. I prefer to develop mine – alone."

"Will you ever perfect your mind?"

"In a million years – possibly."

"Why – what do you mean?"

"The mind never dies. It survives the change we call death and carries on throughout innumerable lives. My object in this present life is to develop the mind as cleanly and industriously as possible in the time allowed me."

"So that mentally you get a flying start, as it were, in the next world."

"Exactly."

"You think each life then is a sort of school."

"Life *is* a school, always has been, and always will be."

These men were not all duffers. It was a surprise occasionally to camp a night with a swagman who could reel off the classics and was familiar with ancient and modern history and with other interests besides.

I also was learning, dimly realizing there was some deep mysterious life going on within and all around us, something controlling the stars, and the air, and the earth, and thoughts and blood, and every blade of grass.

Lamb-marking time found us riding the back paddocks mustering the lambs and ewes into the stock-yard. Then came the drafting, separating the frightened lambs from their agonized mothers. Then came ear-marking, Scotty and I "catching" for the boss. He was greased lightning. Scotty would snatch up a lamb, and sit it helplessly on the rail facing the boss. One "click" and the boss had snipped a piece from its ear; a cut, several bites and jerks and the lamb was a wether; one "crunch" and its tail was cut off as Scotty loosened his fingers and the lamb leaped out into space, its agonized bleat filling the air as it hit the ground. By then the boss was already snipping another victim. All day long, high pressure; shouts of men, skithering of tiny hooves, bleating of sheep, cunning work of sheep-dogs, clouds of dust drying on sweating bodies.

We little realize the work needed and done before our beef and mutton are on our dinner plates.

After the lamb-marking the boss ordered me to ride the boundary fences. Scotty pulled a long face: "Ye'll go loony riding out there on your own."

"I wish he'd sent you with me."

"So do I, laddie; it would be great. But ye wouldna catch me alone out there for all the tea in China. There's enough banshees in this big old lonely wool-shed on windy nights."

"I thought banshees were Irish ones," I grinned.

"So they are, but there's Scotch ones, too. More blood-curdling than the Irish ever knew how to be."

"There's none in Australia anyhow."

"Aren't there! Ye wait an' see!"

Nights were lonely, but it was a great job in the daytime; riding the fences and repairing when necessary, meeting every day a mob of big red 'roos that had their home in a tangled patch of scrub. They used to watch me with the liveliest curiosity. One morning a big red fellow was caught by his leg twisted in the wires. We had a great fight, struggling to get the leg free. Whenever I met him afterwards he would sit back on his tail and stare with the liveliest interest, his big brown eyes looking very wise. We became almost on talking terms.

Every here and there at a corner or angle along the fence, timbered rabbit-pits were dug. Upon them was a trap-door, the rabbit would run along the fence and on to the trap-door, to fall down into the pit. The trap would then swing back, ready for the next victim.

Where rabbits are thick these pits become so full that they can hold no more. At present nature was quietly helping the bunnies for she blew leaves and dried grass against the fence and these sometimes would wedge the trap-door so that it could not open and shut. My job was to clear the pits.

In one pit was a big, black, vicious snake, a score of dead rabbits, and half a dozen live ones huddled on top of one another in a corner. Apparently the angry snake had struck out right and left when he fell into the pit, while the rabbits frantically scuttled all around him. Of the little live group in the corner only their rumps and white tails were visible, wedged into a fear-frozen mass.

In another pit one day the only live things were a snake and a porcupine. Strange company. Each detested the other, yet neither could do the other harm. The porcupine could creep wherever he liked across the pit and the hissing snake must give way. A snake can't argue against a porcupine's quills.

Another pit held a fox, gorged on rabbits. He had tried desperately hard to get out of the trap, rooting and biting at the logs of the pit, leaping at the top logs trying to tear them apart. There was one live fluffy ball wedged deep into a corner of the pit, the one live rabbit, the others he had torn to pieces in maniacal rage. That one live bunny had lived through a night of terror. The red fox crouched in a corner, his eyes like flaming emeralds.

I detested cooking. Making a damper is a misery, washing clothes an abomination. I'd throw my clothes in a waterhole, weight them with stones so that they wouldn't float away, then next day (if I remembered) throw them out on the grass to dry. Many a soddy damper I've baked. Puftaloons are tasty though; fry them in fat, then smother them with treacle and swallow 'em while greasy. You don't notice if they're soddy then.

Nights were a bit lonesome. Some nights brought a faint breeze, just enough to set the leaves whispering. Other nights were so quiet that an insect boring into bark made quite a noise. The screech of an owl sounded piercing on such nights, while a dingo's howl seemed to linger in the air, to fade away very, very slowly. Always, some curious wallaby or bandicoot would rustle in the grass just outside of the fire-light. Sometimes there'd come one quick screech when a native cat grabbed a sleeping bird.

One evening the sun sank angrily behind black clouds, a sighing came from away over the bush. In the quick darkness the tree-tops began swaying then moaning as the wind came howling with rain hissing to drum the tent all night. A half-drowned snake slithered inside but I happened to be lighting the pipe and saw him. Any man will give shelter to wild things on stormy nights – but a black snake looks so poisonous and furtive and slithery when he's half drenched.

Men in the field, Lightning Ridge 1910.

13. LIGHTNING RIDGE

I returned to the station feeling restless, and worth just on fifty pounds now. With care, that sum should allow me to try my luck on the Ridge.

"Better make the plunge and come along too," I suggested to Scotty.

He hesitated, smiled, shook his head: "No laddie, a bird in the hand is worth two in the bush, to me anyway. I like to see it coming in every week. A mickle makes a muckle."

"Never venture, never win."

"True. But I'm not one of the venturing sort. Besides, mining is a rough life, I'm not cut out for it." Poor Scotty, compelled to think of possible ill health.

"Sorry to lose you, Jack," said the boss.

"Good-oh, boss, sorry to leave the station. I've liked the job."

"Oh well, young fellows generally get tired if they stay in the one place too long. There's a job here waiting you if you go broke, any way."

"Thanks, boss."

That afternoon saw me with Matilda up again, turning off the Walgett road on the track past Dungalear station to the Ridge. Black clouds were fast gathering, thunder rumbling, lightning flickering – and fifty miles to walk to the Ridge.

Before sundown the following afternoon came the ring of axes, an occasional halloing of men, smoke rising above the buddha. A big flat partly cleared of timber appeared among low hills, a maze of dumps topped by a forest of windlasses. A few bark huts, tents everywhere, campfires twinkling, the billies boiling, men bending over camp-oven or stew billy.

I camped away quietly that night, and next day bought a tent and tools, billycan, camp-oven and stores. Erected camp near the edge of the timber, then strolled around seeking a bit of spare ground to start a hole.

This was a new step in life. A man had to battle for himself now, think for himself, work for himself. There was no employment here from which a man could draw wages month by month. A man was "on his own". It was a great feeling though, something like what a bird must feel when about to escape from a cage: Shall it fly away and battle for itself into the unknown? Or would it be safer to step back into the security of the cage?

There were miles of spare ground in which to start a hole, anywhere underfoot might lie a small fortune. Almost certainly there would be nothing but hard work. The only way to find out was to try. I sank duffer after duffer. A thousand men were toiling like that, just to "bottom". Slogging down through the sandstone to break through the steel-band on to the opal dirt – the bottom, where opal would or would not be. The hearts of those toilers beat like the thuds of their picks as they dug deeper towards this fateful bottom. The question everlastingly was: "When'll you bottom?"

"To-morrow – the morning after at latest." Men toiled to bottom before sundown lest in the growing dusk their pick break some priceless gem.

"We're right on the band!" the gouger would call up to his mate. "Better knock off," would come the reply; "we'll bottom in the morning."

"Right." And in the sundown they'd trudge back to camp to talk over that bottom by the camp-fire, to dream of it.

Next morning they'd break through the band and expose the first of the bottom. They'd work now as carefully as if picking down into fragile china.

"See anything?" would call his mate, gazing down the shaft.

"Not yet!" would come the muffled reply, echoed by the steady chip, chip, chip of the pick. He'd pick off an inch deep of the opal dirt right across the shaft. Nothing. Send the dirt up in the bucket where his mate in the sunlight would carefully "noodle" it, seeking tell-tale potch and colour that maybe his mate had missed. Nothing. The gouger would dig a foot down right across the shaft. Nothing. Their hopes would fade. When he'd dug down two feet he'd straighten up wearily and call:

"She's a duffer."

"H'm. H'm."

"What'll we do?"

"Oh well, better sink her down another foot or two."

Then they'd abandon the hole and, carrying their tools in the slowness of bitter disappointment, seek about where to start another hole. Immediately their pick sank into the fresh earth, new hopes would be born, and very soon, taking turn after turn in the rapidly deepening shaft they would be working at full speed. As the hole deepened so their dreams would come rising to fever pitch again. *This* time, they might bottom on it.

Each dawn brought a thrill, it was present in the concentrated activity, the tense hope. Before the sun went down, *some* must strike opal. One works well, physically and mentally, when striving for some great prize that lies hidden almost within reach. Soon after dawn the Flat and Gully and Hill would seem to be alive with wallabies that were men walking to their claims. At midday they popped up out of the holes and spread across the Flat towards the camps for the midday meal. At sundown they all came back from their holes again and soon hundreds of fires were twinkling.

Then a restrained excitement, rumours that ripened into whispers, then into certainty. The "Big Four" had "struck it", had "bottomed on large nobbies" (black opal) and looked like developing into a big patch! Lucky beggars! Yes, it was true right enough; they were showing the nobbies now, down at old Ted Dorrington's store.

We'd stroll down. It was thrilling to get a glimpse of the stones. Beautiful things. Then we'd drift back to camp with feverish dreams of bottoming on stones like those.

One particular evening three mates had bottomed on it. Lovely stuff! all pin-points of flame and flash of orange and green. Eager faces appeared in the firelight as we crowded around them. On a sheet of canvas spread on their sapling table lay the pile of black nobbies smeared with opal dirt. The lucky diggers were so eager and excited that they were snipping the stones by light of the fire and a hurricane lamp. These opals in the rough looked like small, utterly valueless darkish-grey stones. The "snips" was a little iron tool with sharp jaws. As the miner's hand gently closed on the snips the jaws came together and bit a chip from the edge of the nobby.

"Ah! Ah! A beauty!"

Under the hurricane lamp blazed a glint of orange and green and red.

How wonderful! The fire and the hurricane lamp were man's own rough light. How immeasurably more beautiful was this light liberated from these dull stones! The light in these stones would never go out, nature had imprisoned it there in utter darkness millions of years ago. And now it blazed in glory.

A miner snipped another nobby. A wave of fire glowed through the dull potch.

"A stone! A stone!"

Carefully he snipped another and another and another. At each snip colour showed. All stones! A parcel! Hundreds of pounds worth of gems lay there – possibly thousands. And how many more lay still buried, down in the just bottomed shaft? Hundreds of stones, thousands of pounds worth of opals. No wonder our nights were dreams of magic stones of fire and orange, of hoards of opals dug from the heart of the earth.

But the Ratters came. Thieves, by night, went down the lucky claim and gutted out the opal. A wave of furious indignation swept the field. If the Ratters had been caught they would have been lynched. Then again they ratted a claim. And the time dawned when only new-chums, or a gouger whose excitement overcame him showed newly dug opals.

The Ratters proved to be an underground gang, jackals of humankind, shadows of the night. Cunning and desperate, they would fight like rats in a trap if caught. And it would be a trap indeed – down in the bowels of the earth.

A scout might be anywhere, in any company, yarning by the camp-fire, mixing with the boys at the store, or on the Flat – anywhere. This "hail fellow well met" type unobtrusively sought to know every one, to mix in for a yarn anywhere. Learning who were on opal, he would report to the Ratters. In the dead of night they would climb down the shaft and work like fiends. Before dawn their look-out man on top would drop a stone down the shaft. The Ratters, with the spoil, would then climb the shaft and separate, each to his own camp, and be in bunk before dawn. The scout, the jackal for the pack, received his cut for the information.

The lucky gougers after breakfast would hurry to work, climb down the shaft and – stand aghast. Ratted! Evidence of maniacal work everywhere: shaft, bottom, and drives mullocked up with opal dirt among which in the candle-flame there glittered spots of fire and orange. Hundreds of pounds worth of opals smashed in the Ratters' mad haste. No one would know how many hundreds, perhaps thousands, were stolen.

The gougers held indignation meetings; sought the Ratters to lynch them. But first – *catch* the Ratters!

For a time I sank pot-holes alone then went mates with little Archie Campbell, a nuggety New Zealander. A great mate, always cheerful. Above all, always willing to cook – but an awful cook. One day he boiled a perfectly good watermelon. Luxuries were scarce on the Ridge. I stared in anguish while the disillusioned Archie tried to fish the pulpy mass out of the billy with a spoon.

"I thought I'd give you a surprise," he explained mournfully. "I didn't know you eat them raw."

Archie and I had no luck. Eventually he was offered the chance of going into a good claim. And I went mates with "Old Tom Peel". We were destined to go through a lot together.

Every one called him Old Tom Peel though he was only about forty-five. Tall and slim, with a questioning face and cheerful personality, Tom had known the vicissitudes of life. Once a rising solicitor in Sydney, an over developed taste in beverages was rumoured to have been the reason for him now being an opal-gouger. We drifted together, a queerly assorted pair.

Matt Watson's camp at Lightning Ridge. Jack Idriess is second from the left and Matt Watson is the seated man with the white beard.

14. TOM PEEL

We pitched camp near the head of the Three Mile, just up the Hill a little. Here, among tall box-trees, the green of the smaller buddha and spurious sandalwood was beginning to show gaps under the axes of the campers – Mick Cullen, Darkie Anderson, old Matt Watson, Nugget Nelson and others of a good-hearted crowd. From here, at knock-off time, we could look down on the activity of the Flat and Three Mile, the red, winding track coming up to us from the bark stores and boarding-house which was now the heart of the field. Cantwells, Nobby's and Sims Hill and Newchums, and now the Three Mile, just found by old Archie Gillespie and developed by Matt Watson and Mick and others, were to boom and produce the most wonderful opals the world has seen.

To me it was a constant wonder what colours far outrivalling the sun and stars, the birds and flowers and butterflies should lie buried in the abysmal blackness of the earth. And that these stones of surpassing loveliness were destined to flash again in the light of day only when manipulated by the hand of man. No other living thing that we know of in our world could have found and appreciated them. Can it be that these lovely things were actually planned and borne in the womb of time to await the coming of man and Eve?

Amidst the excitement of it all Tom and I were busy sinking shafts, then driving tunnels in the. opal dirt, below the sandstone, toiling hard for that elusive opal. While thus driving in opal dirt any blow of the pick might unearth a stone. It was very like pearl-fishing: a pearler might open a thousand pearl-shells and not find one pearl; an opal-gouger might swing his pick ten thousand times and not unearth one opal.

Tom was a great mate, a "we'll-do-it-to-day-if-you-like-but-to-morrow-will-do" type, good-humoured and matey, but one of the worst cooks in the back country. He'd rather live on johnny-cakes and treacle than cook a meal. And so would I. But we both loved the bush, and the sky and the stars at night, and we both had our dreams. It was Tom who actually started me writing. For months he urged it with a persistence entirely contrary to his happy-go-lucky nature. And Mick Cullen encouraged him. I laughed at the absurd idea, thought it was one of Mick's jokes. But night after night, when lying smoking in bunk, Tom would return to the attack.

"You told me the only thing you could do at school was write essays," he'd demand.

"Yes," I'd grin, and blow smoke up at the ridge-pole. Tom, half lying on his side, from his bunk across the tent would wave his pipe in emphasis.

"Then you can write!"

"Rats!"

"So say you, just like a mopoke. 'Croak! Croak! Croak!'"

"You and Mick have put your heads together for a joke."

"There's no joke, we are in dead earnest."

"What have I to write about?"

"All the things you've seen in the bush: the old characters you've met; the interesting birds and animals and snakes we've yarned about. There're hundreds of little interesting things you've mentioned since we've been mates together. The papers would buy that stuff!"

"Rats!"

"Yes, in your head. I know more about that subject than you do young feller me lad."

"Oh talk about something sensible. If I could write about all the subjects in the world, I'd hate doing it because I hate writing I wonder if Scott and Austin will bottom on it to-morrow."

But a night came when Tom started the fuse. A very, very slow-burning fuse that smouldered and spluttered for twenty-five years. Craftily Tom had been yarning about tales in good books for two entrancing hours.

"Those blokes could write," I said. "Those are great yarns!"

"Why don't you do the same?" he said sharply.

I stared, then simply replied: "Tom, you haven't been on the scoot for a month, but you're imagining things ... If I could write, what would I have to write about anyway?"

"Write an article about life on the Ridge."

"That would be easy. But not an editor of all these newspapers you brag about would even consider it."

"You couldn't do it."

"Do you really mean to say I couldn't write a simple essay on life at Lightning Ridge?"

"You heard me."

I laid my pipe on the box where the candle was glued with its own grease, rummaged about for the scribbling-pad and a stub of a pencil. Then wrote rapidly and kept on writing. Tom quietly smoked. The lights in all the distant tents gradually died out, deep night settled down over the sleeping camp. At last it was finished. I tossed the pad to Tom.

"There!"

"That was what I wanted," he grunted. "Good night." He tucked the pad under the old rolled-up coat that was his pillow, blew out the guttering candle, and rolled over.

Next evening Tom pulled out the scribbling-pad.

"Look, Jack, this is good stuff," he said genially. "I'm sure it will be accepted. But just revise it a bit; put in a few corrections, brighten it up a bit here and there."

"I'm finished with it," I growled. "You and Mick only want to show it to the boys."

"Of course we do ... when it's printed. Now come on Jack! You've done the worst. Complete the job, just to please me. I'll post it to the *Sydney Mail* for you. This is just the sort of article they like ..."

I started to revise and became interested. That's the only part of writing I like. Next evening Tom made me go over it again. And the following evening I had to write it all over again in ink – a detestable job. After it was posted I never gave it another thought.

A month later when the mail coach arrived Tom went down for the mail. He returned grinning from ear to ear, and waving a letter.

"Here it is!" he said triumphantly.

"Is what?"

"Open it and see."

Inside was a cheque for three guineas for the article on opal-mining at Lightning Ridge.

Outside Watty Vause's Store, Lightning Ridge, 1910.

15. THE RATTERS

One morning fury broke out, the Ratters had ratted three claims the night before; the owners had found their drives mullocked up, broken opal everywhere. Some thousands of pounds worth of opals had been ruthlessly robbed from the three claims. The owners of one claim had been too excited to keep their good fortune to themselves, but not a soul except the owners of the other two claims knew they were on opal. And yet the Ratters had found out, and also proved their team work by efficiently ratting the three widely separated claims.

The big bullock-bell rang; we all swarmed down to the Flat and crowded around a tall dump. There, a Vigilance Committee was selected and formed, the Committee to make secret plans to wipe out the Ratters. Only the Committee would know the plans, while action was to be secret too. Feeling much more secure the big crowd broke up into talking groups which separated presently to their claims. We felt confident now that the Ratters would soon be laid by the heels.

Next morning the field had settled down to work when Tom excitedly called down the shaft:

"Come up Jack! There's hell! They must have caught the Ratters!"

I climbed up, to hear the deep-toned bell tolling across the Flat and floating away up the Gully. Everywhere heads and shoulders were popping up out of the shafts like rabbits from a giant warren. Within minutes we were all moving down to the Flat, hundreds of excited men converging to form a dense mass around the big dump. Up this was slowly climbing "Bull" Massey, president of the Vigilance Committee. Massey, a shearer when not a gouger, was a short, tremendous nugget of a man with a voice like the bellow of a bull. The deep-toned bell and "Bull" Massey's voice went well together. He was clad only in loose pants and a Jacky Howe flannel. Massey never wore boots; the soles of his feet were as tough as a blackfellow's.

When on top of the dump his upraised hand commanded silence. "Boys!" he bellowed, "two parties here have been jumping one another's claims. It's a dispute as to ground. We've got to settle it."

He then ordered the disputants to line up on either side of him, then bellowed: "Now boys, listen to what the blokes on the right have got to say."

The spokesman of these had his say. Then the spokesman on the left.

When he'd finished Massey bellowed: "Now boys, you've heard what the blokes on the right have to say, an' you've heard what the blokes on the left have to say. Now all those who support the blokes on the right, hold up their hands."

A forest of arms went up. "Carried unanimously!" declared Massey. "The party on the right owns the ground!"

We dispersed to our shafts. It was a quick and simple way of judging all disputes; and the verdict, even in difficult cases was never actively disputed. It was fair because every man on the field could attend if he wanted to and as we all understood our simple rules we could give an intelligent and fair decision. Ticklish cases would be patiently argued out by all the most experienced men, the others of us would be silent.

Old Matt Watson's camp was a centre at night-time. We would stroll over and yarn around the fire.

Little old Matt with his kindly face and twinkling eyes and white beard was a favourite on the field. I doubt if in all his life he ever harmed a man. His memory is fragrant to all old hands of the Ridge. Mick Cullen and Darkie Anderson used to camp beside Matt. Nuggety Mick, with a tantalizing laugh, was an incurable joker; it was a constant wonder that some irate victim did not kill him at least once a week. Darkie Anderson, a small-built but powerful level-headed man, had the priceless gift of laughing at his own hard luck. Behind the veil of the hard years ahead, though, luck was awaiting him, for he was to strike both luck and romance on the field. The romance ended in a wonderful wedding.

All manner of men from all over the field, apart from us "regulars", used to yarn at old Matt's campfire. Old John Landers was a regular. We called him "Old John" then, and after thirty years he is still Old John. John had a tiny income that we vaguely imagined "came from Sydney", an income just large enough to pay his tucker bill. Otherwise he would have had to roll up his swag. He was a man full of good humour and hope. He must have sunk five hundred shafts since he'd been on the field, yet never bottomed on opal. At the Grawon men were bottoming on opal but Old John's shaft was a duffer. His claim was known as "The claim that God forgot". The unluckiest man on the field, he was lucky in that he saw lots of good in his fellow men. In his heart he loved nature, and in his mind stored up memories that were to prove far greater wealth than opals throughout the years that lay ahead.

Andy the Frenchman was another notoriously unlucky one. Andy sank five hundred shafts, toiled for years and years, and never made a rise.

The majority of men were shearers, or gougers from the distant White Cliffs opal-fields. The Ridge was a great shearers' field. Situated about forty miles from the Queensland border, it was the centre of large stations spreading for hundreds of miles around it. The shearers, after cutting out the last shed of the season would make straight for the Ridge on horse, buggy or bike, and settle down there for six months' hard gouging. If they struck it lucky, they would stop. If not, they would pack up and ride away when the new season started only to return with new cheques and hearts brimming with hope.

This black-opal field, practically unknown to Australia apart from a sprinkling of gold- and tin-miners, attracted wanderers from not only such far scattered places as the Amazon, the icy forests of Finland, the oil country of Persia, but also from the most brilliant cities in the world.

These men had histories, some a history which had helped rock governments and kingdoms. Occasionally, by accident, or by some chain of circumstances which had stirred to fever-heat overwrought memories we would get a glimpse of these histories. Perhaps only his solitary Australian mate would hear, listening in the quiet of the tent. Sometimes though a story came pouring out in the sympathetic silence around old Matt's fire. The pasts of other men we learned in broken part as time went on; the pasts of some we never knew.

Steve the Russian we all liked. A splendidly built man with a ready smile, always eager to give a man a hand. Except that he was a Russian and a wanderer we knew nothing about him. He would explain Russia to us when we asked him, though some questions he laughed away. One evening after the mail had arrived we were discussing some Russian news in the papers, bomb outrages, furious riotings. The Cossacks had been turned loose in Moscow. Steve's eyes grew bright in the fire-light, his mouth twitched as he listened. Then Darkie and Kaiser started arguing about Russia and the Czar and Revolutionists and Red Sundays and Russia in general as we very vaguely understood it.

Suddenly Steve sprang up pulling off his shirt as he almost shouted: "I'll show you Russia!" He kicked the fire into flame and stood with his bare back to it. "There is Russia!" he hissed. "Look at it!" His back was terrible. A sea of great weals and ridges of hardened flesh and skin. "The knout did that," he exclaimed through gritted teeth. "And the Cossack arm that wielded it was strong and tireless!"

His handsome face was twitching like a maniac's. He fought to control terrible memories that turned his body into a twitching nightmare. With distorted brow he pulled on his shirt, walked around a while, then sat down by the fire and started yarning. A member of a Secret Society, he had developed into a Nihilist. Was arrested on the very eve of a plot against the Czar's life; was flogged to within a breath of death; suffered hell in prison; then was driven among a herd of humans to Siberia. His story of that terrible march, of life in the mines and prison lumber-camps, the escape and terrible flight across Siberia chased in turn by wolves and Cossacks held us silent far into the night. Of one hundred and twenty men who broke camp he only learned of eight who ever reached safety.

But for us arguing about Russia and conditions there, we would never have known that Steve was a Nihilist, never known about that back of his.

Old John Lander's mate was Bob King, a well-liked man of the then unmistakable English aristocrat army type, with the corners rubbed off. When in the mood he would tell us of episodes in a life lived to the full, colourful and exciting. A lieutenant in Her Majesty's Life Guards he had drifted to China, become a colonel in the Chinese Army and fought in the Boxer Rebellion. He told some dreadful stories of savagery and atrocities there, of Asiatic soldiers mad under gunpowder and steel, of burning towns and ravaged people.

One remarkable thing King had done. Years before, he had written a book called *Hortense: a Tale of the Future.* In it he predicted all about aeroplanes. Privately, we thought airships might come some day in the dim and distant future, but none of us dreamed what we were destined to see. If I had been told I was to see airfleets battling in the skies, that my heart was to stand still at the shriek of falling bombs – well ...

One night the big Guardsman and Old John were camped at the Grawon, their tents a few yards apart. John awoke hearing awful moans. He listened wonderingly – never dreaming of King whose boast was perfect health. Then a pitiful groan raised the last hairs on Old John's pate. He stepped out of his tent and noticed a light in King's tent. King was in great pain; his eyes stared up feverishly, expectantly.

"What on earth is the matter?" asked John.

"Oh – oh – oh! The first pain in my life – an awful one. Oh."

"What is it then?"

"I don't know. Must be sciatica. Oh!"

John sat down on the box, staring with troubled eyes at his mate. King was lying on a bag stretcher, two corn sacks stretched upon poles, the ends of which rested on two logs. Where King sagged down in bunk the bags were barely six inches above the ground.

John did all he could to comfort the stricken man, but in vain. He lit a cigarette, determined to sit by him during the night; then dropped the cigarette, which rolled under the bunk. He stooped to pick it up and saw...

Pressed under the bunk by King's weight was a hefty porcupine, its quills sticking up through the bags. Every time King moved he received a prick from those needle-pointed spines: "Oh! Oh! Oh! I daren't move – I daren't move!"

"Oh well, you won't let me do anything for you and I'm blessed if I'm going to sit here all night and listen to your moaning," declared Old John. "I'm going back to bunk – and sleep." And grinning to himself he stepped out of the tent.

"John! John! You wouldn't leave me?" But hard-hearted John stalked away. Snug in his own bunk he chuckled as now and then agonized groans came from King's tent.

"For heaven's sake stop that howling! A man can't get a wink of sleep!" roared John.

"Oh! Oh! Oh!" moaned King. But as time went on the joke turned against John for he couldn't get to sleep. The joke had gone far enough.

"For heaven's sake, King," he roared, "get out of bunk and look what's under it! Then you won't have any more sciatica."

King, thus adjured, found the seat of his trouble. And the language he used had more points than the porcupine.

16. TRAPPING THE RATTERS

Tom and I toiled on, but fortune refused to smile. Occasionally we joined the noodlers for an hour or two, but only because finances were getting low. This was a wonderful life with fortune anywhere around the corner; I dreaded the thought of having to leave the Ridge. The industrious noodlers were every day like ants scratching at the dumps of the lucky gougers. It was easy thus to tell a claim that was on opal. As each bucket of dirt was hauled up it was dumped, and was then scratched away by the noodlers who, snips in hand, broke and closely examined every clod of dirt. For, no matter how careful the men gouging below were, even though by candle-light they noodled the dirt themselves in the drives before sending it up the shaft, now and then they missed a stone, and it went up in the bucket of dirt.

All stones thus missed were regarded as the legitimate prey of the noodlers. Numbers of men got their start that way. Two arrived with empty tucker-bags one morning and called in at Mrs Hawkins's boarding-house. Jim Saddler gave them a loaf of bread; someone gave them a tin of dripping. They enjoyed a feed, then climbed up on Fred Heath's dump and noodled an £80 stone.

It was *infra dig.* to noodle, though, unless you were nearly broke. And then it was the unwritten law that you should only noodle until you found a stone and could thus fill your tucker-bags. But the majority of men would pack up their swags and go, rather than noodle. When the Ratters became bad, this noodling was a problem. The only thing to do was to stack the dirt in the drives then, when the patch was worked out, to send it up the shaft.

Cobar Mary enlivened the field now and then. Mary was a determined black gin, a giantess. She had walked all the way from White Cliffs to follow up the opal-fields. A very sophisticated lady of stern mien was Cobar Mary. One Monday morning she appeared waddling up the red track across the Flat. She carried a rifle and a steely glint in her eye. She mounted a dump.

"Git aht of here!" she commanded.

"Why? What's wrong, Mary?" asked the startled windlass man as he backed away from the rifle.

"You stand back nah, or I'll blow daylight through you! Gahn! back down that dump!" And he did.

Mary lifted her voice in a shout that turned the heads of every windlass man within hundreds of yards.

"Below!" she yelled down the shaft, "you yaller livered codfish! Below thar!"

The gouger toiling below looked up in alarm.

"Yes, you! you crow eatin' tief, Joe —. Send up the money you owes me for washin'. You dirty big double-crosser!"

"Go to hell, Mary!" shouted the angry man. "I'll pay you when I knock off and not at all if you come this sort of game on me!"

"Youse'll pay me now! Right now! Send up that five bob in the bucket before I drills a hole through you!" She pointed the rifle down the shaft only to spring back and poke the muzzle in the windlass man's face.

"You would would yer!" she screamed. "Give a woman a bang on the head when her back's turned. Now back straight down that dump!" Furiously she forced him down the dump, her big black face contorted in fury. By this time hundreds of men were climbing up out of shafts to watch the fun.

Again Mary pointed the rifle down the shaft. "Send it up before I shoots."

He stared up at the savage black face glaring down the shaft. We all knew Mary. Call her bluff and she would respond – instantly. She cared for nothing, she was just brute instinct. She would shoot right enough. Nothing surer.

"I haven't got a shilling on me, Mary," he called up. "You should know that very well."

"Send up your cloes!"

"But Mary—!"

"Send 'em up!"

Slowly he put his flannel in the bucket, reluctantly he unbelted his trousers.

"Them boots too!"

Mary manned the windlass with a wary eye on the windlass man away on the other side of the dump, and hauled up the bucket. Then she went through the clothes.

"Jest as empty as yer head, an' that's full o' bugs. Climb up."

"Send me down my clothes."

She kicked a rock down the shaft. "Some more's comin' down," she howled. "Climb up!"

He commenced to climb while Mary rolled the clothes into a bundle and kept them under her arm. Presently, the gouger's head appeared. He implored for his clothes, Mary had him where she wanted him; she told him a lot of things all about himself in a voice that bellowed over the field.

Then she picked up junks of opal dirt.

"Come right up an' let 'em see you before I fill you in with rocks."

There was nothing else for it; furiously he climbed up out of the shaft. But Mary held the rifle-muzzle straight at him, while her eyes blazed and insults poured from her lips. It was obvious that her excitement alone might touch off that trigger any second.

He started walking down the dump, heading in the direction of his camp across the Flat. And as he walked Mary's voice bellowed all over the field advertising his beauty points.

And thus Mary for all time ensured prompt payment for "washin' done".

It was not the last time she came stalking among the shafts with a rifle, for she had various misunderstandings with gougers at odd times. She invariably evened matters up by waiting until the victim was sinking a new shaft. Catch a man down a shaft and he is helpless – it doesn't even need a gun. Mary could bring the bravest man to reason simply by throwing stones down the shaft.

The only other aboriginal on the field at that time was Tom Bowen, a different character altogether. A fine type was Tom, a good man, cheery, willing, well liked by all. A wonderful specimen physically, standing six feet six inches. He could fight like a threshing-machine too. He was a cobber of Bob Kirkpatrick's. In years to come, Bob and Tom went to France. Bob came back, Tom stayed there.

There was another great old coloured character on the field though, "Old Black Joe". Old Black Joe had strayed far from "de ole plantation". A short, squat, wrinkly-eyed old American negro, he had been a renowned American prize-fighter in his prime. Now he was a respectable gouger who one day frightened the life out of a young fellow. Both were driving, tunnelling down below. Each man heard the thud of a pick directly in front of him, coming towards him. Old Black Joe listened. Then: his eyes developed a nasty gleam, his wrinkled old face and bald head sweated in bovine anger. Ratters! ratting his claim, and in broad daylight! He'd show them!

Old Joe wasn't on opal, and he didn't stop to ask himself how a Ratter could rat his claim while working in another. Presently, he could hear not only the deep thud of the trespassing pick, but the wrenching tear as each blow levered out a lump of opal dirt. Then he heard the lumps falling to every thud of the pick. He put all his hard old muscles into it and worked like a demon.

A lad was working on the opposite side and wondering who it could be driving towards him. Suddenly, a pick point burst through the wall within an inch of the lad's face. Then a bestial grunt, a shattering blow that knocked opal dirt all over him, a candle-flame, and an apparition glaring at him through the hole in the wall, a black, sweating, mad-eyed apparition slavering from gummy jaws.

"What you doin' dar!" roared Joe.

Blackness as the candle went out for Old Joe had thrust his head through the hole. A howl of rage as his shoulders caught, mad grunts of rage as he swung the pick to widen the hole, then grunts from the heavy body as it squeezed through the darkness.

The young fellow fled. The next morning, with definite news of Ratters, a thrill ran through the field: "A Ratter has been blinded by a dynamite cap! Another has had his arm blown off!"

Throughout the day the matter was discussed from camp to camp, from shaft to shaft. No definite particulars could be nailed down, it was only one swiftly spreading whisper that two claims had been ratted in the night. In one claim the gougers had buried dynamite caps in the face. A Ratter's pick had struck a cap which exploded.

The gougers owning the other ratted claim had embedded a cap in a plug of dynamite, craftily bored a hole in the face, inserted the plug, then smeared it over with powdered opal dirt. The Ratter's pick had struck the cap and in the blinding explosion the pick was wrenched from his hand, his arm being shattered.

But what had happened since? Who owned the two claims? Where were the stricken Ratters? Very few of us ever definitely knew. The details were kept secret by the gougers concerned lest desperate vengeance come. And the Ratters kept quiet lest the mob go mad on a manhunt. It was whispered later that by night they had driven the hurt men the fifty miles to Walgett, then another thirty miles down the line and entrained for Sydney.

For a few weeks that shock curbed the Ratters. But only until their espionage system had caught up with the latest methods to catch them. All the secret traps set by the Vigilance Committee, by gouger bands, and by individual mates proved again in vain. One night a whispering rumour spread like wildfire throughout the field. That afternoon there had been a phenomenal find of opal at a well-known claim. It was true right enough; the lucky gougers could be seen in shadow within their lighted tent, examining a rich pile of nobbies.

But the Vigilance Committee who all night lay around that claim, waited in vain. The Ratters never came.

Trap after trap was laid. A claim would strike opal. The lucky gougers flashed the gems. It was well known they hated working night shift. The candle would burn in their tent late at night; they could be seen turning into bunk. The candle would be blown out and then, an hour later if any night owl cared to peer into the tent he would surely hear the men breathing, could touch them.

Away out on the dark diggings, away down below in the inky blackness of the tunnels, men lay hidden in the rich claim – waiting. But the Ratters never came, their espionage system was perfect.

Yet with uncanny instinct they still located the lucky claims – where no new traps were set. They overcame the dynamite cap menace by digging into the toe of the face, then gouging upwards to bring down the layer of opal dirt under the roof. It was this layer which carried the opal, and the caps, if any. Once the first six inches of this layer was down they could work like fury again.

Few gougers cared to risk planting dynamite caps in the face; the risk of picking them out again was too great.

Thus the ceaseless war between the gougers and Ratters went on. The only way in which the lucky gougers could be sure of their hard won riches was to sleep as well as work down their claim when on opal.

Sharpening picks, Lightning Ridge 1912.

17. BLACK OPAL

Old "Doc" Hendrick was a general favourite, though no one knew or cared whether he was qualified or not. A great old chap, he had saved many a life in the back country. A giant of twenty stone, he towered above us all. With his huge chest free to the sun, his sleeves rolled up, his broad, determined face ready to scold or laugh, he was a tower of energy and strength. He sank a very occasional shaft to "keep his weight down", but the field in general "looked after the doc". There was no other within fifty miles.

Naturally, Doc Hendrick had no real surgery, only a few medicines and few if any tools. But he possessed the experience, initiative and confidence to tackle any job. Some years before this, old Tom Rowe was thrown from his horse, and his skull badly fractured. He was taken to a township where several doctors gave him three days to live. His people sent for Doc Hendrick. Old Doc lumbered across to the blacksmith and directed him in the making of a clamp, and various tools. With these, somehow or other, he forced the man's skull to open out, and patched up the fractured bone. Then he put the clamp on, and screwed until the skull-bones came into position again, and presently old Tom regained consciousness. (He died only a few years ago aged ninety-three, so the cure was perfect.) That was the sort of "handy man" the old Doc was. He had successfully patched up numerous broken humans.

One sunny day Darkie Anderson (not *our* Darkie) and his mate went to their shaft to collect their tools. It was a forty-foot shaft, a duffer. His mate lowered Anderson down on the rope. This did not quite reach the shaft bottom. Anderson was leisurely tying the tools to the rope when he felt "it" getting him. He snatched up at the rope and shouted.

"Pull up quick! Foul air."

His mate bent his back to the windlass as he had never done before. Anderson's face, tight set, and with teeth clenched, came staring up the shaft. He clung desperately to the rope. Immediately his head appeared above the shaft the fresh air got him, his fingers slipped and he crashed straight down. His mate yelled and young Ross came running. Ross gamely went straight down the shaft and held his breath while desperately tying the rope around the unconscious man in his arms. But the gas caught Ross before he could knot the short rope around himself.

"Pull up!" he shouted, and clung to the rope. He'd only gone a few feet when he too fell back.

Benny McGillick and his long mate Flannigan came running to the shouts of the windlass man. Long Mick MacCormack and Mick Moran came running too. As they swung Anderson clear of the shaft they were unknotting the rope. One whirl of the windlass and the rope whizzed back down the shaft.

"Hold the rope!" said McGillick. He took a deep breath, seemed to jump at the rope, and whizzed down the shaft to a sound like blazing paper – the skin was burned from his hands.

He snatched up Ross, whipped the rope around him and rapidly knotted it; he felt the gas ... there was not time ...

"Quick!" he shouted and clung as man and rope seemed to leap up the shaft.

"Quick! hold my legs!" said Flannigan, and his arms and head and shoulders were down the shaft. MacCormack and Moran each grasped an ankle and clung, with Flannigan's long body head first hanging down the shaft. He knew what would happen immediately his mate's head came up to the fresh air.

When Ross appeared, McGillick was still clinging to the rope, teeth clenched, eyes bulging. Flannigan's arms swung out deep down the shaft and McGillick seemed to just slide into them – unconscious.

Anderson was badly hurt, his legs smashed in several places. He was in great pain when Doc brought him to. The old Doc made a wonderful job of those legs. In the final setting he stood away back while Bryant, staring at him, manipulated the legs.

"Steady!" said Doc, "steady!"

His hand was held out in a straight up and down line with his eyes, and the hand and voice and nodding head gave directions to Bryant as carefully he worked the fractures back into place.

"No ... a little below the knee ... No ... Yes ... Just there. Steady now. Now the ankle. No ... a shade *this* way ... That's right. Steady now!"

Later, after Anderson had been driven the fifty miles to Collarenebri, the hospital doctor examined him: "Splendid!" he said finally, "a perfect job, With a complete hospital at my command I could not have made a better job than that."

So it was no wonder that we, and many a far-out bush township thought a very great deal of the old Doc. He did me a good turn later on. But not because of sickness or accident.

Tom and I toiled on, but couldn't strike opal. And yet it was being found all around us, Opal King after King sprang up.

Long after Bob Kirkpatrick and Ernie Marr struck it rich. They were among the first of the real Opal Kings; no one ever knew what wealth really came out of their wonderful claim. Percy Marks bought thousands of pounds worth of gems from them to take over to the Franco-British Exhibition. At least three hundred gougers must have been on opal. Some among them while their luck was in made hundreds of pounds per week.

"We must have killed Chinamen, or looked cross-eyed at a black cat," growled Tom.

"How'd you like to be an Opal King?" I asked.

"It'd please me better than wearing an emperor's crown," grinned Tom, "instead of this old hat with no crown at all."

We were walking back to camp at sundown after a hard day's toil. Before us trudged Big Billy and Little Billy. Great mates these two, always together, great team workers in mine or camp. Very quiet men.

"Their luck's out too," I nodded. "There's hundreds of us."

Big Billy threw down his tools at the forge and bent to light the fire. "Get a billy of water," he growled to his mate.

"Get it yourself!" growled Little Billy.

Tom and I winked as we walked by.

"I got it at dinner-time," expostulated Big Billy.

"Well, get it agen," growled Little Billy, and squatted down.

"Go to hell!" said Big Billy.

Little Billy slowly arose. They walked towards one another, fire in their eyes. They got into it. A willing go. Skin and hair flying in quick time. The big fellow put it all over the little fellow until Little Billy got his second wind. Then he pasted the stuffing out of Big Billy.

"Hold hard!" gasped Big Billy at last.

"Had enough!" gasped Little Billy.

"Yeh."

"Shake!" And they shook, clung gasping to one another's hands and shoulders, then woefully grinned around at the crowd.

"Well, boys, the show is over," Little Billy panted.

We all laughed, and drifted back to our camps while Big Billy picked up the billy-can and trudged away for water.

That evening at old Matt Watson's camp several lucky gougers brought a parcel of stones in the rough for Mick and Darkie to value. As they turned them slowly round and round we'd get a flash of fire and orange where the stones had been snipped. Valuing was work in which often an expert was deceived. The job was to try and estimate which stones would "face". When the rough potch was ground off the stone by the emery wheel, would the colours show unblemished? If so, what colours? Red, flame, orange, green, gold, or a mixture of all, and other colours too? Would it face a fire stone, a pattern, a pinpoint, a flame, or a peacock? Would it prove a prized harlequin? Would it be a bar stone? What would be the depth of opal? What brilliancy of colour or flash? Or would the stone be sandspotted, or milky, or cloudy, or simply dull and lifeless?

Every stone in the rough was a fascinating puzzle! It might be worth £100, it might be worth five shillings. How then, to form an estimate of the value of each stone? And the value of the parcel as a whole?

During the first few years the stones were invariably sold in the rough. A buyer might give £200 for a parcel, and sell it again for £500, perhaps far more. It would be resold in England and America eventually for thousands. On the other hand, that parcel might face "sandspotty" and not be worth £50. But experience taught both gouger and buyer.

Now, however, the nobbies were gradually being faced, that is, the rough was ground off them and the beautiful stones polished, then sold on their face value.

Macintosh was perhaps the first cutter, Charlie Deane came soon afterwards; and others gradually came. Eventually, most of the gougers owned their own machines, merely a dentist's lathe. But sure judgment was necessary in facing the stone. One turn of the wheel too much could spoil a £100 stone. Several turns too few, and a stone worth £100 would only appear to have the value of £50. Thousands of pounds worth of opals thus imperfectly faced were sold, to be bought over again by speculating buyers and gougers, refaced, and resold. Indeed, numerous valuable stones were resold many times. Some noted stones, originally sold for £5, have been refaced, resold and speculated with again and again to an eventual sale of hundreds of pounds.

An illuminating instance on a grand scale was the Queen Alexandra, a beautiful stone found some time later. It was one of the great jewels of the world. Every Australian has heard of the Koh-i-noor yet not one in 200,000 has ever heard of the Queen Alexandra, the Light of the World, the Pride of Australia, the Flame Queen, the Pandora, and other world-famous gems unearthed in our own Australia. McNicoll unearthed the Pandora after my time on the Ridge, refused £1000 for it on the field. Goodness knows what it would be worth overseas. One claim sold their opals for £13,000, four years later they estimated that the parcels would have realized £30,000. Charlie Dunstan unearthed a beautiful stone that he sold for £100. It was resold and resold overseas, disappearing and reappearing and disappearing as the world's great gems invariably do. At long last there appeared a photograph of a stone in a publication of the

Washington State Bureau, America. It was identified by men who had handled Dunstan's stone as being the identical same. Whether it was or not, I couldn't swear to, but if so its value was different, it was now marked as being worth 255,000 dollars.

One of the famous stones was sold for £450. It was exhibited in a Sydney jeweller's window at a price above £2000. Another was sold for £92, and before it left Australia £2000 was refused for it.

The Queen Alexandra was a famous instance where one turn of the wheel too much, or one turn too little meant all the difference in a fortune. It was first sold in the rough, according to rumour, for £30. Then resold for £75. Eventually it was faced. It proved to have a false "face"; a light smoky scum dimmed beautiful colours underneath. The problem was whether the grinding of this scum away would cut out the colour which was only of cigarette paper thickness. Eventually, someone took the risk. And the stone blazed in glory.

Rumour at last drifted back to the field that this stone went to the Durbar Exhibition and was finally sold for £30,000. But it was seldom that such figures could be really traced. Once a noted gem stone or a first-class parcel left the field it disappeared from us to begin its wanderings through many hands and into far-away countries.

But all these rises in prices, the rapidly growing appreciation of black opals in America, London, Paris, Berlin, Vienna, was only just developing. And we were gaining experience in valuing and facing and the various things to be learned about opals, particularly the black opal, then new to the world.

Thus, on the field itself, it was a fascinating game, buying and selling opal. An old treacle-tin full of potch and colour that a man could buy for thirty shillings might "face up" worth £5 worth of opal. It might even contain a £10 stone. Lots of men got their start going around the camps buying potch and colour, then classing and facing it; to sell it to buyers, or speculative gougers who again would sell it at a profit, and so on.

As the field developed and values gradually began to be "sensed", various gougers with the "colour sense" tried their hands at opal-buying – a small parcel for a start. One such would face it and, if his judgment proved correct, make a profit on the parcel by selling the faced stones to one of the regular, or visiting buyers. Flushed with success, eagerly he would walk around the field seeking to buy a larger parcel. Quite a number of men were successful at buying; a few quickly developed into big buyers and eventually travelled the world selling the gems. Percy Marks for instance. He was toiling now as a gouger, climbing from his shaft caked in opal dirt. He was destined to become a great buyer, to leave a trail of opals around the world. He was to be awarded the Grand Prize for gems at the Panama Pacific Exposition, to dazzle American eyes with Australian opals, to be awarded the Grand Prix at the Franco-British Exhibition, and to receive an honour from France after the great display at the Lyons Exposition. Those opals caused a sensation in America and France. But these events were in the future. Daily we saw Percy Marks swinging his pick but we little dreamed the brilliant future awaiting him.

There must have been at least three score of wanderers who had carried their swags to an opal-field and had never seen opal before; but who through success in buying saw London and New York, Paris, Berlin, and Venice, Constantinople and Moscow, and the great cities of India. Men who rose to sell their gems to rajahs and emperors and millionaires, to count and duke and lord and merchant prince. Thus romance hovered over the opal-fields for every man – and not in the finding of opals alone.

At the start of Lightning Ridge there was a recognized opal valuer, Tommy Lewis. A direct eyed, swarthy man with a heavy dark moustache. To him all parcels were taken to be .classed and valued. He charged a shilling in the pound. He'd value a parcel at say £350.

"Ask £400," he'd growl. "Take £350."

And if the gouger received £350 or thereabouts then Tommy received his corresponding shillings. Many scores of thousands of pounds worth of opal passed between the slim fingers and before the dark eyes of Tommy Lewis. But time and experience are quick teachers. Enthusiastic gougers found that they too could value opal. Like the monopoly that the very necessary and enterprising early cutters enjoyed, the monopoly of Tommy Lewis gradually passed away.

Tommy, like most nomads on that field, had led an adventurous career. He once said good-bye to life in the thunder of a howling storm, a terrible shipwreck in which he was one of the very few survivors. His valuing on the Ridge brought him easily enough to retire on. But years later he was driving a drover's cart in south-western Queensland.

The busiest resident buyer on the field was E. F. Murphy, a big, quiet, black-bearded man with a slow smile. Perhaps the eyes of no human in the world have seen such visions of loveliness as have flashed and danced and glowed before the eyes of Murphy. One of the pioneers and earliest buyers at the famous White Cliffs light-opal field, he had come to the Ridge soon after it broke out. Then for years the brightest opal gems the world has ever produced passed before his eyes. If those incomparable lovelinesses from the greatest beauty in the heart of nature could have only been preserved to posterity in some form of a picture screen!

Murphy bought for Wollaston, the far-seeing man who, after years of heart-breaking struggle, put Australian opal on the world's market. He followed the Queensland prospectors over terrible country; his own mate perished of thirst – within half a mile of water. When, after years of battling, prospectors found the Little Wonder Mine then Wollaston knew that opal such as the world had never known before, was discovered. But years of struggle and bitter disappointment yet lay ahead of him in wandering among the markets of the world before he at last placed Australian opal on the map. Meantime, away out at a place destined to be called White Cliffs, some thirty miles from Wilcannia in New South Wales, kangaroo-shooters had found queer stones with a little colour in them. Wollaston was quickly on the scent.

From 1891 to 1903 White Cliffs officially produced more than £1,500,000 of opal; the unrecorded sales are unknown. The millions that this opal realized overseas is unknown. Australia never realized the value of her gems. We sold for hundreds of thousands, what was worth millions. For years Australia produced a gem whose like was previously unknown to the world. Then, when the world appreciated its value, we didn't. Even to-day, Australia owns the only black-opal field in the world.

When Charlie Nettleton found Lightning Ridge, a marvel unknown was uncovered. Black opal. The world refused to believe it. White and light opal yes. America in particular had "gone wild" over Australian light opals during these last few years. But black opals! No. Such a thing could not be.

For two years the energetic Wollaston could hardly sell black opal anywhere in the world. For three years it was sold only in the rough. Then, again, America began to buy. That was just when the first opal-cutters came on the Ridge. Suddenly America, then the world, went "mad" over black opal. The world bought, bought, bought, and Australia again sold for hundreds of thousands of pounds gems really worth millions.

But Old Tom Peel and I, digging in the very heart of the only black-opal field in the world, could not find a solitary stone. We were slaving now, toiling to strike opal before the last of our scanty money was spent.

18. STEALING GRASS

Kaiser, like others of the mystery men, intrigued us. He was a grim type of man not given to confidences. We would have seen little of him had he not taken a strong liking to Nugget Nelson, a mate of Mick Cullen's. Nelson and Maori were perhaps the only two men on the field that the Kaiser really liked. Kaiser was a short, rather dark man with a straight shut mouth and determined chin. Defiant eyed, gruff and somewhat overbearing, his rather Prussian manner had immediately earned him the name of Kaiser. We thought him German until one night sentiment touched him and he spoke wonderfully of Sweden, and of moonlight over the Norwegian fjords. In the course of time probably he confided more to Nugget Nelson than to any other man in Australia. His father, he told Nugget, owned a shipyard on the Tyne. But he went under the name of Charles Anderson, because he had to have a name.

"No one knows my name!" he declared fiercely. "No one ever will."

"One of the Legion of the Damned," murmured Tom.

I glanced at Tom. Flickering fire-light was shadowing memories across his face.

"I knew a man," said old Tom Watson, across the fire, "who lost himself as completely as Kaiser. But when he died they found a birthmark. He was traced from that."

"I too," laughed Kaiser bitterly, "have a birthmark. But when I die I hope the dingoes pick my bones."

Fateful words.

He had left White Cliffs with £1800, a beautiful pair of horses and buggy. He arrived on the Ridge with only what he stood up in. Old Sid Sharkey stuck to him for £60 worth of stores. Kaiser pitched camp at the Four Mile, toiled like a fiend, and eventually struck opal. Out of the first parcel he put aside the best stone, a little gem. He sold the parcel, paid Sharkey his £60, then posted the gem to Sydney to be mounted with diamonds as a tie-pin.

It was a magnificent pin. He presented it to Sharkey.

Kaiser was now building a camp down on the Flat, a maze of sheds and passages covered with box-tree boughs, destined to become notorious as "The Bungalow". Things happened there that would not be believed if I wrote about them. Kaiser was subject to mad drinking-bouts. But would suddenly sober up, turn furiously on his drinking friends, clear the camp, and not touch a drop for fairly long periods. One of the Legion of the

Damned, his story was a mystery. And he was to live through grim chapters yet before the wild dogs picked his bones.

Another chapter had started for me. Nearly broke, it was a case of again shouldering Matilda and looking for a job.

"I'm coming back," I swore to Tom. "I'm going to strike opal yet."

That was the lure that pulled us back to the Ridge. Every week men quietly faded away, packed horses or bike or themselves and disappeared. They'd reappear again in six months' time, "chequed up" and smiling.

That night at old Matt's "Council" strangers appeared. The "Pony Drover", and his men. A quiet man was the Pony Drover, with the sharp glance of a quick thinker, a constant traveller. One learns to instinctively realize the jobs of some men.

"Do you want a job, Jack?" smiled Mick Cullen.

"Yes."

The Pony Drover was staring at me.

"I want a horse-tailer," he said quietly.

"Right."

"Very well. I leave in the morning."

"I'll have my swag rolled."

"Good."

So away I rode with the Pony Drover. It was a great little outfit, known far and wide in the back country. All the horses were nuggety little grey ponies, the hardiest, most intelligent, cutest little team that ever stole a cocky's grass in all the north-west. In his droving trips, whenever possible, he followed the country shows. He entered the ponies for various ring events. They were great hurdle jumpers and trick ponies;. balancing on barrels, drinking a bottle of beer and winking their thanks, leaping through hoops of fire, standing to "beg", waltzing to music or whip-up to every circus trick. At bush work they were cunning as sheep-dogs. And they would live on the "smell of an oil rag" if they had to, and still keep their condition. But they didn't have to – that was my job. Not only was I horse-tailer, I was to be grass-stealer.

The horse-tailer in a droving outfit is responsible for the horses. He *must* keep them in good condition, otherwise the whole outfit collapses. The travelling mob whether sheep or cattle may unavoidably be starving; in drought-time beasts may be dropping out and perishing day by day; but if the horses starve, then the outfit comes to a standstill and the whole mob may be lost.

We travelled in easy stages down through Walgett and Burren Junction to Narrabri.

Narrabri in those days was a straggling bush township, different indeed to the prosperous, modern town of to-day. Now there are large stores, good streets, electric lights, motor-cars. Then, it was a dusty road with straggling little stores and houses, and horses hitched up to the rail before the pub. The pub inside was remarkable. Walls and ceilings were painted with really artistic paintings of famous homesteads. The district was different too. Then it was sheep and bush; now it is up-to-date mixed farms, sheep, and a famous wheat-belt area. The bush has practically disappeared.

Strange, the coincidences, chances, recurrences, or whatever you call that .which links the milestones of a man's life. Many years later I was to revisit modern Narrabri, and speak in the Town Hall at the Agricultural Conference. If any one had told the horse-tailer to the Pony Drover that a day would come ...

When I was leaving the hall a grizzled old-timer held out a freckled paw.

"Hey," he grunted, "so you're the young tag that uster thieve the cocky's grass! I was a cocky them days an' one night you put your whole hungry team on a little patch of clover I was savin' for my prize cows. If only I could have caught you I would have tanned your hide so you could have skinned yourself for boot leather."

"You would have had to get up early those days," I laughed. "But another cove tanned my hide all the same."

From Narrabri we rode on down to Boggabri and soon afterwards the Pony Drover lifted a mob of five thousand wethers from Baan Baa station. We started crawling north then, the sheep were to be delivered to a station near St George, across the Queensland border.

The Stock Route, for considerable distances, was eaten bare as a bone by large mobs of travelling sheep. Some of those sheep were so skinny you could read a newspaper through them – as the saying was. Water was plentiful, and there was an abundance of grass – in the stations and cocky's paddocks. It must have been torture to the hungry sheep, slowly mooching on day by day living on a fallen leaf here, an overlooked root there while on both sides of the Stock Route but within the fences was sweet, green grass.

But the sheep were the drover's concern. The horses were mine.

Every drover knows that the most suspicious man in the world is the

Man on the Land – when travelling stock are about. Every station would send a stockman to ride with us when travelling through station country, to pass us through as quickly as possible and make sure we stole no grass.

If travelling through a selection then there was sure to be a stern, bearded horseman hovering on our flanks or a cute, slouch-hatted, bare-legged lad. Or some tomboy of a cocky's daughter would be shepherding us, riding parallel away among the trees, her shrewd suspicious eyes watching our every move. By daylight at least there was never any chance to slip up the fence wires and drive the sheep through for a stolen bellyful.

And at night on camp we knew very well that the mob would be watched by someone out in the darkness. It was the same on another trip when I was droving down the Castlereagh. It is the same everywhere. All eyes watch the poor drover.

But – they cannot see far on dark nights! If you can dodge their ears, you win. We'd camp at sundown, wherever Cooky had prepared camp. I'd take the ponies half a mile or so back along the route, hobble and bell them, then return to camp and feed up. There'd be a man on watch; the rest of us would yarn around the fire. We could hear the horse-bells tinkle just now and then; so could any one else, if any one was listening.

Drovers go to bye-bye early. Whether yarning or under the blankets, I'd be listening intuitively, lest any horse-bell or hobble thief was busy. There was no need to actually keep listening. The faintest unusual tinkle, or an unusual silence would instinctively warn me, even, at times, if asleep. But there was seldom cause for uneasiness until the camp was sound asleep. And only occasionally then, of course.

Still, along every Stock Route was sure to be some local bell-and-hobble thief; some lad or man who spied on hobbled horses then, when the drovers were asleep, crawled out and softly, very carefully gained the confidence of the horses, slipped the bells off their necks, then the hobbles from their legs, and disappeared quiet as a shadow. These occasional marauders bothered me not because of the bells and hobbles they might steal, for they rarely got the chance. The ponies might be out there now but soon they'd be several miles away, a marauder couldn't find them. But while I was also on private business, I might just stumble on a hobble thief. That happened one pitch dark night. Cautiously walking towards the ponies, I tripped right over a hobble thief crawling towards them. We up on the instant, trembling in the fright of our lives. He vanished as I hurled a stick at him.

When the stars wake up the drovers go to bye-bye; when the stars go to bed the drovers wake up. The stars don't use a "hipper" but the drover does if the ground be hard. He scoops out a hollow for his hip-bone then spreads out the blanket, rolls up the coat as a pillow and pulls the blanket around him. The fire burns down.

My great friend from Lightning Ridge days,
Jack Allsop who rode with me in the Light Horse

Then I'd slip away. The ponies' bells would be silent – they knew. I'd take each bell off and tongue it; slip the hobbles while they stood motionless. A hiss, and they'd be quietly walking back along the Stock Route.

Perhaps for two or three miles, to some likely portion of the fence I'd noted while riding behind the drovers by day; some place where good grass was just inside the fence and where the wire was apparently a little slack. Carefully I'd lean down on the top and second top wire over the centre panel and weight them down. Generally they proved too taut, because certain suspicious people always keep their fences lining a Stock Route well repaired and the wires invariably tightly strained.

But if the top wires were slack at all they could be forced down; then I'd pull a wire hook from my belt and hook these top wires to the bottom wires. Thus the ponies could step over the wires. They waited patiently, then quietly in turn stepped high over the wires into the sweetly grassed paddock. But if the wires weren't slack I'd have to walk to the strainer post where the ends of the long wires are strained. With a key (wire straining key) I'd quickly untwist the two top wires and pull the ends through the post. The ponies would then step over the three bottom wires.

Now, a horse-tailer stealing grass, if he be a good man, leaves as few traces as possible. He must leave tracks of course, but what the eye doesn't see ... That wire, if possible, must be strained back and left all shipshape in the morning. You can rob the squatter or cocky of his grass if you are smart enough to get away with it, but you do not leave fences cut or down. It simply isn't done.

So when the ponies were munching the grass I'd make all ready for the morning, and a quick get away. It generally was not possible to pull the already strained wire back through the post-hole and strain it, because having once been strained it was too short. So I always carried a small coil of wire, cut the strained end off the fence wire and joined a short piece of wire to it by a figure eight knot. Then slipped the wires back in place and took a hitch around the strainer-post.

In the morning well before dawn I'd be back, slip loose the wires, then round up the ponies. They'd never stray far, and always stuck together. They'd move quietly straight back to the exact spot in the fence, step over one by one and wait while I quickly strained the wire with a small wire-strainer, quick and easy tools to work with. Then, I riding bareback, swiftly we'd make towards camp, arriving just as Cooky's fire started gleaming up from the greying earth.

If a suspicious cocky rode up just at dawn and watched the drovers turning out for breakfast he might stare at the ponies' poddy bellies and wonder just where they had dined. Even if he rode back along the fences he would find no wires cut – everything would appear in order; he and the drover and the horse-tailer would have an easy mind.

19. THE RUINED HUT

Of all the grass-stealing feats during that long, dry, hard trip, the taking of the ponies through a cocky's front gate and past his veranda to feed them in his own horse-paddock, was the limit. It was a newly-built house fronting the Stock Route; a double gate in the stock fence opened on the drive to the house. A nice house. Perhaps this selector was the first in this district of those keen, thoughtful men who saw money and prosperity in coming closer settlement. His paddocks fronting the Stock Route were dry and bare, probably overstocked, so as to obviate the worry of "broken" fences and remove temptation from passing drovers. A double temptation, for just here was a long muddy waterhole, a recognized camping-ground of every travelling mob that passed this way. No doubt this selector's back paddocks were well grassed.

Anyway, his little horse-paddock adjoining the back yard was six inches deep in sweet grass. But even drovers would never dream of raiding a paddock that began at a man's own back veranda. The cocky could sleep untroubled at night even though the camp-fires of drovers, with thousands of starving stock burned barely three hundred yards from his front door. He watched our fire while sitting on his veranda puffing his pipe after the sunset meal.

When the night slumbered the boss and I put the "sneakers" on the ponies. These were padded slippers which slipped easily over the hoof and with one buckle were fastened tight. We only used these "pony moccasins" on occasions like this. The boss must give me a hand with the gate; I'd noticed it was fastened with a new-fangled padlock which might defy our spare keys. No use cutting the wires either; not only because of dogs, but I'd noticed a bullock-bell cunningly attached to a wire. And there were sure to be other tell-tale bells; probably one led to this shrewd cocky's own bedroom. Pull a wire and a bell would jangle the night awake. Then out would pour the outraged family with guns and dogs.

Oh yes, this prosperous little home was the cocky's castle; it was impregnable. Only one small thing had been overlooked – the gate hinges. Historians tell us that if only you look long enough and carefully enough and patiently enough you can find a way into the most impregnable fortress. And this cocky had slipped with his gate hinges; they were of a type from which the gate could be lifted off the gate-posts.

The boss must give me a hand to lift off that long double gate. "All set Jack?" he murmured.

"All set, boss."

"Right. Let's get going."

The night was pitch dark, the grey ponies a blur as they padded noiselessly towards the gate. Silently we tried our keys in the padlock. Useless. Quickly we greased the hinges so they would not squeak, then the boss got a grip on one end of the gate, I on the other. Gently and steadily we heaved up, lifted the gate right off the hinges, carried it straight forward, then to the side, and laid it gently down. The rest was up to me. I touched the halter on the leading pony, the others followed in single file. We walked around well out to the side of the house. I'd memorized a route free of wheelbarrows and beehives and kennels as we were riding past during the afternoon. Noiselessly we turned in around the back of the house, then on into the back paddock. Not a dog barked, not a fowl cackled.

I bent to take the sneakers off and hobble the ponies, the greenhide hobbles without any tinkling links or chain. In this little horse-paddock the ponies couldn't wander far but they must be quickly found well before daylight. Cautiously I returned to my virtuous blanket.

Well before daylight I brought the ponies out, their bellies distended like bladders, the boss was waiting like a black post at the gate. Silently we carried the gate back and fitted it on to its hinges. Then followed the ponies to camp.

That morning just as the drovers were ready to move off the cocky came riding out from the homestead, puffing an early morning pipe. He rode beside the boss and yarned as the sheep moved off. He had not the faintest suspicion, though his bushman's eye noted the ponies.

"Your ponies are in great condition," he drawled. "Must be expensive, carting feed."

"Yes," drawled the boss as the mob moved off.

The grizzled old cook, loading his cart, winked solemnly at me.

Pleasant days, peaceful nights. For me, it was a gruff "Jack!" from Cooky well before the first streak of dawn. An effort then to roll out from the blanket, feel for the bridle and yawn away out in the dark seeking the ponies.

Bitter cold when the frost was on the ground, icy finger tips thrust deep in pockets, toes cold as the eggs in ice-chests. The trees felt the cold as well as I did, their still frozen leaves and cold bark told plainly they were wishing for the sun. The men woke with the earliest sleeping bird, crawled to Cooky's fire, lit up and risked a hurried wash. Then squatted around the fire hungrily getting outside a warm breakfast. The sheep all lying there, thousands of them in little misty mobs huddled together and gazing towards the fire. As I would come trotting up with the horses crimson rays were splashing up over distant trees, cart and camp slowly emerging from the grey dawn. The men would stand up, draining the last of their pannikins. As I hurried for a wash and breakfast by the fire, they'd slowly light their pipes, puff up, then reach for bridles and stroll to their ponies.

The more energetic or hungriest of the sheep would be rising, singly, then in twos and threes moving off, watched by the dogs around the fire. Soon, the whole mob would be gradually moving off. The men very slowly riding, or walking, leading their ponies. The boss might stay a moment to yarn with Cooky, giving locality for that night's camp, then he too would follow the mob. Cooky and I would light up, yarn awhile, then pack the cart with the swags and gear. Then the cook would drive off while I rode on after the mob.

Quiet days, just dawdling along. Each stage only to travel a few miles, the object being to allow each sheep every chance to get as much as possible into his belly so that he would have strength to travel another day.

Some drovers will only "follow cattle". With them the movement is a bit faster, the bush not so closely settled, the trips longer, and there is more incident in the less settled country. Some trips carry excitement too especially with a wild mob. Charging steers, "rushes" for water, mad stampedes in moonlight with the mob thundering through ghostly bush.

Terrible thunder that, elemental, something terrifying in its roaring speed. It awakens an unreasoning fear immediately responding to the outbreak of uncontrolled life gone berserk; the primal terror maddening the beasts coldly grips the reasoning men. No man who has ever screamed forward in a bayonet charge, or who has ever galloped with his hair on end in the heart of a mad stampede, will ever forget it.

One night I got a fright that was not caused by any stampede. It was a bleak, wispy, moonlight night. I was driving the ponies back past an old deserted hut near where good grass grew in a station paddock. The ruined walls of the hut loomed up and the ponies suddenly stopped.

"Ssh! Gee-up you cows!" I hissed.

But they stood there. I reached out and smacked the plump rump of the hindmost. He didn't move, and I felt him trembling under my hand. I was about to smack him again ... but stood there listening, staring. Utter silence. The ponies and I like living statues, staring. An icy touch made clammy the back of my neck. Suddenly the ponies wheeled and galloped madly away unheeding my half-choked shout. Open mouthed, with hair rising on end and heart thumping, I slowly backed away, staring at the old ruined hut. I daren't turn my back to it. I stepped back a long distance before hurrying after the ponies ... They were at camp huddled around the fire, the boss with them.

"I don't know what came over them, boss," I panted. "We got to the old ruined hut when suddenly they turned and bolted. Nothing could have stopped them."

"It's all right, Jack," he said quietly. "Better turn in. They'll have to go hungry tonight."

"No fear. I'll drive them back again."

"No hope. They won't face that way again."

"Why not?"

"I'll tell you in the morning. Better turn in. You can't do anything tonight." And he walked to his blankets.

Next morning he mounted his pony and was about to ride after the mob without saying anything. I stepped up to him:

"Boss, why did the ponies bolt?"

"There was a murder at that hut years ago, Jack."

I stared at him: "Do you mean to say that the ponies saw ...?"

"Yes, Jack. Other men have tried to camp near that hut but their horses always bolt."

And he rode away.

20. THE STOCKWHIP

Once our sheep "settled down" we just let them drift along, feeding as they went, about five miles per day, or from water to water. Sometimes a stage, of necessity, was a little longer. They must be "nursed" lest the weaker ones drop out and become food for the crows. Each such tragedy meant a loss to drover and owner. The drover had contracted to deliver the mob in good condition at only a very low percentage of losses. And it is the drover's pride to deliver his mob with the barest possible loss.

One morning the mob moved off, but eighty sheep lay stretched on the ground. The boss examined them, then said: "Poison weed!"

Along well-travelled Stock Routes, patches of poison-weed country are travelled quickly across. But occasionally this weed will grow up in places where it was unknown before. And hungry sheep will eat it. I've seen a thousand sheep stretched dead, poisoned in a night. Perhaps they had picked up the weed throughout the day.

Cattle sometimes suffer similarly from poison weed and vine, and bush. The horse can suffer too. Even the hardy camel can be laid low by a certain deadly bush. There probably is sound sense in the bushman's idea that the goanna eats a certain poison weed as an antidote to snake-bite. When the two fight, should the bitten goanna prove victor he will waddle with all speed to chew a certain weed. Bushmen believe that this weed contains poison which kills the snake poison.

Generally we'd split the mob, let them dawdle along in two mobs about half a mile apart. Thus they spread out over the country and individually had a better chance of picking up a scanty bite here and there. They must not reach camp and water until sundown. Then, with their bellies reasonably full and topped off with water they would camp contentedly. There would be little need of a watch at night.

On hot summer days, though, when towards sundown we drew near water, they had to be held back, necessitating patient work by men and dogs. When nearing water the leaders would begin to lift a nose and sniff. Presently, they'd begin to step out, a bleat would break out here and there, and the thirstiest would get ready to make a rush for water. We'd head them to steady the leaders down.

If the waterhole ahead was small and boggy we'd hold the main mob back, only allowing them to drink in very small mobs. Otherwise the thousands would just simply rush and pile up on top of one another in the waterhole. Besides being bogged, many would be smothered.

shoulder, the swag down along the back. Thus the weight being evenly distributed is not felt nearly so much. If you don't pack your swag scientifically the shoulder-strap will rasp into your bones and gall them like a badly fitting saddle on a horse.

I had only hoofed it about twenty ·miles when I met an old sundowner also carrying an "angle". Easy to see at a distance he was a "professional". Besides the angle, he carried in one hand a large billy-can with puppies' heads poking from it; the other hand carried a water-bag. Corks dangled in a bobbing circle around his hat brim to keep the flies away. Smoking peacefully, he was travelling at a steady pace "makin' for the Castlereagh".

The true sundowner travelled a regular "run", stopping at selected homesteads and towns for food. He was a professional, unlike the author, who was just a casual "humping bluey".

These old professionals carry (or did then) as many as four sets of billy-cans, one set neatly fitting within the other, all with lids on too. The old fellows are very particular about their cooking tools; it's not the etiquette of the road to boil a plum duff or cook a stew in the tea billy. I discarded etiquette; it's easy enough to swill out the billy after you've cooked stew in it; the tea isn't too greasy. It's bad enough having to carry one small billy-can for tea anyway. I'd boil meat in the tea billy if I wanted a change from chucking it on the coals, or I'd boil a doughboy in the billy

There *have been* disasters that way, both with sheep and cattle. It is terrible to see a mob of thirst-maddened cattle, out of all control, rushing a waterhole to pile up on top of one another in a frightful, seething mass.

Droving grows on a man. The quiet life, the moving forward day by day. You get to be reconciled to the winters, the misery of occasional rains, of wet camps. Lots of men are drovers all their lives. But a man has to "grow" into the routine. For me it was far too quiet, the dawdling day by day too slow. But every week meant another twenty-five shillings to go into the tobacco-box for tucker at the Ridge.

This outfit were seasoned drovers, each man steady and quiet. I doubt if anything could have flurried them, from the boss to the cook. They were no wasters of words either. Dreaming along on horse or foot by day, a quiet yarn at night, then each man pulled his blanket around him with saddle and bridle beside him. And the mopokes took over the night.

The only real excitement was a thrill – for me. I didn't like it. Thus dawned my Waterloo.

I'd noted a panel of this particular fence that appeared loosely strained. And that night found it actually so. An easy job. Just weighted down on the top wires and hooked them to the bottom. The ponies stepped across disdainfully. I hobbled them and strolled back to camp. And no one saw but an owl.

Before daylight I returned for the ponies. Couldn't find them immediately – nor even when the trees took shape. I became anxious when the sky lightened. The ponies had strayed. It was a large paddock and it was after sunrise before I located the ponies. The inevitable rogue amongst them had broken his hobbles and enticed the others away. Hurriedly I undid the hobbles, strapped them round the ponies' necks, mounted, and started them back for the fence at the canter. Birds were chiacking one another, a breeze came murmuring through the bush. As we neared the fence I rode easily, considerably relieved, and turned in the saddle for a final glance around.

Away back among the trees a man on a big grey horse was just bending over his horse's neck, breaking into a gallop.

I dug my heels into the pony and drove him on top of the others. They bounded forward and in seconds were going all out; I blessed their circus pranks that made them alert on the instant. The thud of hooves sounded ominously close, the long legs of the big grey were travelling faster than the ponies'. I glanced behind. The pursuer was swinging a long stockwhip – a certain purpose in his face tingled my hide. I snatched at a light branch; it snapped off and I held it. The fence in plain sight now, the ponies galloped in single file knowing we were in for it.

I heard the shivery swing of the whip, glanced round while riding full tilt then, when the long lash came hissing, thrust that branch behind me. The long, greenhide thong bit around it, but its tip caught my ribs like a red-hot wire. As the lash struck I'd twisted the branch, desperately seeking to entangle the thong and thus jerk the whip from his grasp. The manoeuvre almost won out. I stared back at his surprised eyes as he swung back the whip-handle to feel the thong caught, lashed around the branch.

But he was an adept and rode the horse well behind with knees and rein while swinging his body and long arm to disentangle the whip. The thong tore loose and there was a glint in his eye as he swung the long whip far behind him then round and round in the air. He was entering into the zest of the thing – didn't think of the fence.

Down came the whip, the twisting, snake-like thing coming straight to suddenly coil under like a figure S. And it caught me and the pony across our tails like a sizzling slash from a red-hot wire. My mind flashed a picture of the back of Steve the Russian. He struck three times in rapid succession but only partly caught me. That branch was a heavenly foil. I hope no one else has ever sweated in such a duel. To use that whip he must keep his long-paced horse nearly twenty feet behind the galloping pony, but man and horse were expert even though the pony was no slacker; he'd tasted the whip and jumped and twisted like the acrobat he was to save his hide.

The devil behind now struck trying to curl the thong around my branch then jerk it away but I whipped the branch down and with a waving motion captured the thong. He jerked back on the whip-handle manoeuvring it at all angles as an angry fisherman does whose line is caught in a snag. But I was retaliating similarly with the branch, and now that his every attention was occupied made the most of the opportunity by swerving the pony aside. He followed naturally as the pony zigzagged, but I was zigzagging to give the ponies ahead a chance to clear the fence otherwise we would pile up on the fence in a mob.

He flicked the whip away, then half standing in his stirrups brought it right back and high over his shoulder – and down. The pony and I jumped as one, and jumped again and yet again. Just see how high you leap when urged higher still by the bite of a greenhide whip. One agonized glance showed me the first pony taking the fence with the others at his heels like a little mob of 'roos. The zigzag manoeuvres had now brought us fifty yards behind the last pony. It was time to go.

I flung the branch straight back into the grey horse's face. It swerved and I was over the neck of the pony who was flying. As the last pony leaped the fence my heart almost stopped for the pony's hooves touched the wires and they swung up into place. But the lash came down and lifted me and the pony and thus we took the fence.

Thank heaven, the pony just cleared it, it was the jump of his life even though he had cleared many a showground hurdle. My tail felt like raw steak; I felt I could jump the moon let alone a five-wired fence.

Lightning Ridge looking South.

21. THE SUNDOWNERS

When we galloped to camp with me clinging to the pony's mane, the boss's impassive face told he realized just why I couldn't sit down. It is all in a horse-tailer's life. The men had started the mob, the sun was well up.

"Better saddle up straight away Jack, and ride, or walk on," advised the boss. "Carry your breakfast with you. We'll pick you up at camp to-night. Cooky will give you a bottle of oil, you can treat yourself when you get along the road a bit."

"Right-oh, boss. I feel I'll need a *bucket* of oil."

"You'll be tender for a day or two," he smiled, "but it wears off."

"This won't," I protested. "That whip bit *right in!*"

He nodded. "Treat it with oil, and keep the flies off for a day or two. It'll be all right."

"I dunno, boss," I said miserably. "If I've got to crawl about on tiptoes brushing the flies away with a switch, I'm in for a bad time."

Holding the old trousers well back with one hand I gingerly mounted and, half standing in the stirrups, rode carefully and miserably away. No one can understand the pain left by a greenhide whip unless he actually feels it.

The wires bouncing back into place had stopped the devil on the grey horse, of course. If he intended to follow up and have a row he'd have to canter two miles back to a gate, then canter along the Route until he caught up with the drovers. The boss would stolidly know nothing about it. Of course those were his grey ponies! Well, and if they had been in a station or a cocky's paddock they had no right to be there. But he knew nothing about it. That was the horse-tailer's job.

And where was the horse-tailer? Where? Oh, where? Far away behind a clump of trees gingerly treating his tail with oil. The pony gazed sympathetically on, indulging in a fellow feeling twitch now and then. At least he had a tail to keep the flies away.

When the droving trip was finished my cheque was big enough to keep me in tucker for nearly twelve months. So I rolled Bluey at the angle, and started the long tramp back to the Ridge.

The angle swag is neatly rolled (mine never was) with the weight evenly distributed throughout. It is kept tight by two straps, each one about a foot from the end. To one of these straps is joined the shoulder-strap to which is joined the tucker-bag. On starting out, the tucker-bag should balance the swag. The tucker-bag thus balances in front of one but I'd never go to the trouble of cooking a plum duff, it's quite enough trouble having to boil the billy.

Some of the old sundowners of the old Murrumbidgee Whaler brotherhood, those who "followed the rivers" all their lives, were characters. Proud of being sundowners, their method of life meant nothing derogatory to them; they were as "good as the squatter" or any other man for the matter of that. They lived their own life in their own recognized world and had evolved their accepted code of laws and conduct. Not every man who carried a swag was eligible for the League of Sundowners – not by a jolly long way. We "casuals" always felt embarrassed when in the society of these Knights of the Road. We merely "humped Bluey". They merely tolerated us for the time being, if our company was unavoidable. Some of the old "greasybacks" wouldn't even speak to us.

In addition to their own code and society and unwritten laws, they had their secret symbols, recognizable by all the fraternity. Before he reached a station homestead the real sundowner knew whether it was worth while calling in or not. Some hidden sign by the track told him all he wanted to know. Similarly with a township. It might be a "good" town, or a place to avoid. The character of butcher, baker, police, or other notables was known to the sundowner before he entered the town. The information would be all there ready for him "under the bridge". It was not only the police who had their "Secret Service Bureau".

Among themselves, the sundowners had to mind their p's and q's. They had their "run" upon which no other must unduly trespass. And little things counted. It was "God help the cow who's gone an' hung his billy on the same tree as I did twelve months before".

That is not so ridiculous as it appears. The old chaps following a river carefully chose their camping-grounds, and on travelling back up the river would often camp in the same spot as they had the year before, build the fire in the same handy place, bake the damper in the same possy, shelter under the same old tree, fish in the same waterhole. A "good camp" held much the same significance to these old-timers as home does to us.

If a stranger happened to be camped there then there were glowering looks indeed, a glum, "don't speak to me!" night both for sundowner and trespasser.

One sunset I witnessed a furious row in a bend of the Barwon not far from Boorima station. Farther along, a sundowner was just building up his fire, singing:

> Once there was a swag-man camped on a Bil-la-bong –
> In the shade of a Cool-a-bah tree! –
> Up jumped a jumbuck to drink at the wat-er hole –
> Up jumped the swag-gie to grab him with glee! –

Another old sundowner blew along; strolled on to his old camping-ground – and saw the stranger in possession. The sundowner glared around and then:

"Who broke that limb down?" he demanded, "I've hung my billy there once every twelve months these last twenty years back!"

The ensuing argument nearly ended in murder. Fortunately I was a witness. It is certain that similar trivial episodes *have* sometimes ended in murder in lonely parts of the bush.

There are various styles in swags. The collar swag is worn straight up and down. In the rolling, which is very important in every swag, the contents in the blanket are neatly and evenly distributed the full length of the cover. When the swag is lightly rolled, an envelope is made of the cover. The swag ends are then covered in like a roly-poly pudding. It is carried over one shoulder and arm.

Many men carried a small tent-fly as the outer swag covering. Thus, if it came on to rain, they rigged the fly and were quite independent of the weather. Camped by a waterhole, they fished until the weather cleared up.

Invariably they carried several books in their swag to help pass time away, and as "food for the mind". They swapped these books with one another when meeting along the track, and at station homesteads. Hence, a real sundowner had command of a library, and in his well-regulated life did not care whether it hailed or shone.

Travelling in lonely bush by no means meant a dearth of news to them. They had their own "bush telegraph", and passed on news from one to the other. The distances and rapidity with which the news spread was wonderful. Friends could keep in touch with one another though travelling hundreds of miles apart.

Cooking on the roads is a well thought out science. The meals these men could dish up, remembering that everything they used they must procure and carry, is hardly believable. But to go into details of the cooking part of a sundowner's life, of what he can and does prepare out in the open bush, would require several chapters. Most of them have (or had) their own medicines, and mentally treasured recipes for sicknesses, snake-bite, etc. Their little lives were very full and, what is more, the real sundowner was happy.

I determined to discard padding the hoof as soon as possible. Travelling with a swag was too slow. But it was to mean a lot of battling before I rose to the speedy eminence of a horse.

Padding the hoof down the road, musing on lines of speed one hot morning, I heard shouting, and coming around a bend in the track saw an old swaggie laying down the law, waving his arms to emphasize his points, scornful fury in his voice. His swag, listening in an attitude of cocksure disdain by the billy, was propped up against a stump beside the track. The swag was the audience, the audience was Australia.

"When I'm boss cocky of this gorforsaken country," he howled, "I'll drive youse like a shepherd drives a mob o' sheep! You're only jumbucks. You ain't even that, you carn't find yer own eatin' grass! Yer so dumb yer allows yerself ter be fleeced an' they don't even paddick yer fer it! What do you know about the rights o' Man? Yer dumb goats! About the Guvernmint? About politics? Nothin'! Nothin'! You ain't got it in you! You're bone from the shoulders up. You swaller all wot them pollyticians tell yer like a baby swallers mother's milk! It all goes down the same way an' wot don't go out stays in! Yer eats with yer bellies but yer never uses yer heads, only to scratch 'em. An' yer wonder why yer fingers is full er splinters! Put *me* in Parlyment, yer dumb cows an' I'll put you on top o' the world! Every man'll have a job, every man'll work six hours a day, every man'll own a bank. Ho! An' what will the wimmin own? Yer wanter know! They'll own the wash-tub they washes in, you fool. An' I'll give 'em all votes too. All you can give 'em is babies. Wen I'm Premier —"

I coughed, while gazing away across the track. The future Premier collapsed, strolled across and sat down by his swag, pulled out his pipe and studiously began to cut tobacco. He gave no sign, no invitation, so I carried on down the long, hot road.

22. THE FRIGHT

Fortune smiled. Although the season was nearly finished I picked up a job as rouseabout at a shearing-shed, one of the best unskilled jobs of the bush.

The shearers, highly skilled men, toiled on the boards at terrific pressure of muscle, sinew and nerve. All was speed; the hum of the machines, the tense attitudes of the human machines, the constant fall of the snowy fleece, the rumble of massed sheep in the pens, the shouts of the rouseabouts, the push and slide of the shorn sheep down the chutes, the agile horsemen, the snappy work of the dogs – all was concentrated efficiency in the human machine running at full speed – speed – speed.

The "gun shearer" there was a two hundred-a-day man, the tally of the lowest was over a hundred a day. The shearers worked by contract at thirty shillings the hundred; the shed-hands and rouseabouts at thirty shillings per week and tucker.

The shearers tuckered together and, of course, did themselves well, paying their own cook and offsider. The cook was a well-paid man indeed and master of all. Well he knew it. The rouseabouts' tucker was much better than station tucker. I lined my skinny ribs at that shed, and laughed quietly when the rouseabouts complained about the tucker. It is always so; I've done it myself. Have come into a shed ribbed like a poor goanna, then a month later, with ribs well lined, growled at the tucker.

This hive of activity by day, the bustle of meal-hours, the crowded quarters by night was a startling contrast to the quiet life of the bush.

At knock-off time came the noisy rush for a wash and meal, the shouting and laughter, the jokes and concertina and sing-songs and boisterous relief after the strain of the day. But sleep quickly drugged us into bunk to gather energy for the coming dawn.

Sunday was a welcome resting day; we loafed and read and lazed. I'd cobbered up with a big yellow goanna that was constantly sneaking around the hut on the look-out for what he could pilfer. A wicked looking old go', as cunning as a circus flea. One Sunday morning he came waddling around the hut to stare with his long neck held high above stubby front legs. He licked his jaws with long black tongue, his bright, beady eyes staring at an ants' nest. It was one of those big hard mounds, swarming with large ants. Topping the nest were three large emu eggs, placed there by a shearer who had paid an Abo. stockman to carve the shells. The eggs were bad, the shearer had pierced each shell so that the ants could swarm inside and eat out the dead emu chicks.

The go' coveted those titbits. Cautiously he waddled to the base of the ant-bed, then suddenly ran up the bed and violently butted an egg with his nose. As the egg rolled down the ant-hill the go' leaped aside like a cat on hot bricks to roll over and over in the grass violently rubbing off the clinging ants. Hissing viciously he waddled back to the egg, now well clear of the nest but swarming with large, outraged ants pouring out from inside. The go' rushed the egg again and tumbled it over as he leaped aside. Thus more ants were rolled and brushed off against the grass. Again the go' rushed and butted the egg, and yet again, as a man might push a burning barrel with his hand.

I wondered how the go' would break that big thick shell and dispossess the ants from the meat inside. He butted it towards a stump, stood back, surveyed the position, then carefully butted it to within an inch of the stump. Then stood side on and with a violent blow from his tail knocked it against the stump. The shell broke in four pieces. He thrust in his snout and delicately seizing the dead chick threw it into the air, then wiped the ants from his lips by vigorously brushing his jaws against the ground. He waddled to the dead chicken and threw it into the air again. Each time he got rid of a few more ants. Then he seized the chicken and violently rubbed it against the ground. Thus he rid it of all ants. Leering at me, he licked his chops, then guzzled the chicken. He'd earned the titbit.

When the shed was cut out there was a whirl of movement; horses being run in, buckboards loaded, swags strapped to bikes, movement everywhere as all hands packed up, mounted, and rode away, hurrying to the next shed, fifty miles distant.

I packed the cheque in the tobacco tin, threw the old goanna a final bone, rolled the swag, and cheerfully set off again for the Ridge. An occasional wallaby was mate enough; I whistled to the minahs and they whistled back shrilly, warning other birds of possible danger coming.

Afternoon shadows came, followed soon by the silence of evening. I camped by a lonely gilgai hole. The fire burned down; from a nearby bush came a snappy little bark as a cat-bird made remarks about the human. I hissed and twittered in inviting friendship to the "furry" feathered little bloke, but he hid his slaty body in the bush, "barked" disdainfully, and was silent. I turned in and fell asleep, to wake up with hair standing on end. Awful subterranean groans, developing into gasping roars trembling into woeful screech after screech. Silence. Then – screeching splutters growing into roars again.

Trembling in the blanket I stared out into the night, then snatched a fire-stick and poked the fire together. Those deep, agonized roars, those terrible screeches broke out again. Leaping up, I glared around, ready to race for life. An inkling of what the thing might be curbed my trembling senses. But this sounded too awful, too monstrous; like some demoniac thing tortured in an agonizing death-struggle with – what?

Those roars now quavered through the bush; the thing must surely burst asunder.

With a fire-stick clubbed in one hand and a torch of blazing bark in the other I tiptoed out into the darkness. Yes! There it was! A big toad-frog being swallowed by a snake!

The maddened snake had bungled the job; had struck just a second too late. The frog had blown itself up as the snout gripped. The snake's eyes bulged like gleaming black beads; fiendishly they glistened under the wavering torch. Its jaws were hideously distended as it struggled to swallow the toad; it writhed up and forced its weight upon the bloated thing trying to press it into the ground, to lever its own jaws wider and engulf it.

But the toad was blown out to fantastic proportions liable to burst any second and blow the snake's head off. How the bladder-like thing retained sense and a fighting movement was a miracle of nature; but it did. The snake was struggling to swallow a tough bladder energetically jerking and floating and bouncing from its jaws. And just didn't that toad roar! Like a bull! And screeched like a bursting throttle.

In relief after the fright I thumped the snake. The liberated toad bounced right over my head, I had to search around to find it. I kicked it lightly, and it sailed over the bushes like a kid's balloon.

I went back to the fire and sat down and had a smoke, wondering what a snake would feel like if he swallowed a big toad-frog and it blew itself up inside. But of course it couldn't; the snake would puncture it with its teeth when it was swallowing it.

23. THE MONARO COCKY

I arrived back at the Ridge full of beans. Many more tents showed among the trees; bark huts were springing up too. It already seemed a far day from the first tents, then the great event of Mrs Kirkpatrick's boarding-house. The Three Mile was beginning to look like a mighty rabbit-warren. New Opal Kings had arisen, and hundreds of men had "struck it". There were several new stores, half a dozen opal-cutters, four resident buyers, and visiting buyers coming from Europe. The Ridge was booming. A dozen or more women were on the field. Even a travelling circus. It was good to see the boys' smiling faces again.

Tom Peel was working with a fellow gouger; they were on potch and colour.

"We expect to strike it any day now, Jack," he said excitedly. "My luck's going to change at last."

"I hope so, Tom. I'd like to see you an Opal King."

"No such luck," he grinned longingly. "But we've got every chance of striking a patch."

So for the time being I went gouging on my own. Climbed down Jeff's old claim, lit the candle, and stared around.

The drives were all mullocked up, nothing here but silence and mildewed air, the clammy gloom of an abandoned mine. Still adhering to the sandstone roof was a hard layer of opal dirt. I wondered that they had not picked down that opal dirt. Perhaps they had got their opal below it, and considered it waste of time to pick down that valueless layer of rock. Right at the shaft-mouth stood the crumbling remains of a small pillar of opal dirt, left as a support for the roof. Near the roof the pillar was barely nine inches thick. Setting the candle in the mullock I began chipping at this pillar, using the little driving pick. At the third chip there sounded a wee "crickly" noise, like the blade edge of the pick gently scraping glass.

With a sudden thrill I held the candle to the pillar. There, in the dull grey earth gleamed a speck of colour, that flamed as the candle slowly moved before it. My heard thumped; this was a stone!

Carefully I gouged around and under, then pried it out with the spider point. Then held it under the shaft-mouth and turned it around and around. Where the pick had scraped it, glowed a spot of dull red. In smiling excitement I gently snipped the opposite edge of the nobby. Brilliant green flashed in the candle-light.

I laughed and laughed. A stone! The first stone I'd ever dug.

Putting the precious thing deep down in my trouser-pocket, I, crouched by the pillar, chip-chip-chipping. How gently a man's hands wield the chisel-pointed pick when he's – on opal!

"Click!" I laid down the pick, chuckled to the candle, then held its light to the face. A speck of brilliant orange danced from the dull opal dirt.

Another stone! I sat back and laughed up at the sky framed high past the mouth of the shaft. Stars twinkled up there in the daylight sky, but none shone so brightly as the tiny stone in the face. Tenderly I gouged it out. It was the same size as the other, and a band of orange ran right through it.

Another stone lay buried behind it. Just the three. Then the pillar was pierced through, leaving only the vacancy of worked out ground behind it. It was thrilling examining these stones, staring around into the darkness of mullocked-up ground, then at that seam of opal dirt clinging to the roof. If it, too, only held opal!

Walking back to camp for the midday meal was glorious. There was opal in my pocket. Surely the birds had never sung so sweetly.

Only old Matt and Mick and Darkie and Nugget and Tom knew about it. Not another soul, until I'd worked out that dirt still left on the roof. I worked it out in a week, worse luck. The three stones from the pillar realized £5 each, £15 before dinner. I'd have had to work four months on a station for that amount. From the roof I chipped out £80 worth of opal. In a week made almost £100. Two years' work on a station!

The feeling of independence was greater than £10,000 would have brought to many another man.

"Your luck's changed," said Tom.

It was a bright moonlight evening; we were walking past old Ted Dorrington's store where the usual crowd of gougers and a few women sat yarning, when a wild yell startled us. Sailor was walking down the track, stark naked.

"Mad!" murmured Tom. "Not drunk!" The men grabbed Sailor, took him across to Gadsby's and tried to put trousers on him. He resisted violently: "The Lord never wore trousers," he swore, "neither will I."

Poor Sailor. An old man-o'-war man, he loved porcupines; swore they were the best dish he'd ever tasted. When not gouging he was always walking the bush searching for them. This night he had set out to walk through the bush and start a new shaft. But he had sunk his last shaft.

"I've an idea Canada Bill will go the same way," said Gus Monk.

"Watch Canada – he is queer." Old Canada did go that way: Suddenly tackled old John Landers at Gooraway with a knife. It was a sudden, desperate struggle.

In a couple of months Tom's claim duffered out.

"It's only a stringer, Jack," he said dolefully. "Potch and colour that strings us on and on, like a siren luring the sailors to the rocks. I'm born under an unlucky star – I'll never strike opal."

"Tommyrot. When you've duffered your claim right out we'll go in mates again, we'll strike it some day."

"The claim is duffered out now. I'm fed up and so is my mate. He's pulling out. We don't know who's the Jonah. What if we start a new shaft to-morrow? Your luck seems to have changed."

"Right-oh."

So we started work again but the old luck still held. We couldn't strike it though opal claims were all around us. Opal was rising in price too, its value being a little more appreciated. Even so, few of us realized that a parcel sold then for £100 would probably be worth £300 on the field in two or three years time – and a thousand overseas.

"Opal all around us," sighed Tom, "and not a stone in sight. We must have done something – like Jimmy the Murderer."

Jimmy the Murderer was still scowling around the field. Jimmy had an awful face. He once owned a barrel-organ and a monkey, the pride of his heart. It had all a monkey's tricks. Jimmy would leave it chained to the end of his bunk when he went to work. The monkey would open the safe, and take out a crust of bread. Then, with a bag, sneak out of the tent to the end of the chain. He'd cover himself with the bag when he heard the Happy Jacks coming. This squawking, quarrelling, gossipy crowd of birds visited all the camps when the men went to work. Under cover of the bag, the monkey would throw out a fistful of crumbs and the Happy Jacks would come gobbling them up. Like lightning a tiny brown paw would snatch out and a Happy Jack would disappear under the bag. The wretched monkey would quietly screw its head off, pluck it, then throw it out. Then scatter more bread-crumbs.

When Jimmy returned for the midday meal plucked Happy Jacks would be lying about the camp. But the monkey would be coiled asleep under the bunk.

This cute monkey loved beer, which was its downfall. And Jimmy's too. The monkey insisted on drinking his beer from a bottle. One of the lads put a packet of salts in the bottle and the furious monkey went off beer for good. Jimmy would have earned his nickname could he have caught the joker. The monkey, now off the beer, refused to act, and Jimmy, in a fury, killed him. Hence, Jimmy the Murderer. And he'd had no luck since.

Several small circuses visited the field. They did a roaring business; but finding conditions so congenial gradually merged into the life of the field. Each circus had three or four girls – and jolly good types they were too. Naturally, all the young half-axes were intrigued. It was the usual thing of evenings for them to stroll into camp and suggest:

"Who's coming down to see old Fred?" But it wasn't the proprietor they wished to see. "Old Fred" was a very popular man when he brought the circus to the field.

The girls quickly laughed their way into favour; they could easily handle the young half-axes. But some of the sterner males became badly smitten. One hardcase sat up all night writing his love letter. It was a masterpiece; he showed it to us with pride. The last sentence read: "I'm not offering you the dregs of a waster's life, or the drainings of an empty pocket." We congratulated the author, and awaited results ... It was a laugh.

Old Fred eventually sold the circus tent to the Social Club, and the celebration over the deal led to the most surprising fight ever seen on the field. A popular young fellow was the "Monaro Cocky", a nuggety, clear-faced lad with an engaging smile. His dad worked a selection in far-away Monaro.

On the field at this time had congregated a little fraternity of "half-pugs", who mostly camped and worked together. When the celebration in the circus tent was at its height a half-pug knocked out the Monaro

Cocky; struck him swiftly and entirely unexpectedly. When he came to he looked around in a dazed way, stood up, and drawled to a friend standing by:

"Did you see the man who knocked me?"

"Yes."

"Who was he?"

"Oh ... never mind about it, it's all over now."

"Who was he?"

"Better forget about it."

"Who was he?"

But the man would not tell, he did not want to see the Monaro Cocky get knocked about. So the lad from Monaro went quietly back to his camp and slept.

Next morning early he was at his friend's tent. "Who hit me?"

"Forget it."

"That man knocked me when I was unprepared. Now, I am going to knock him. Who is he?"

"Oh – very well then. He was one of the half-pugs. You can find out for yourself which one."

The Monaro Cocky walked across to the half-pugs' camp. They were sitting outside, just commencing a drink of tea.

"Morning."

"Morning."

"One of you blokes knocked me last night. Who was he?"

A man jumped up. "I did."

"Very well. Now I'm going to knock you. But first I'm going to have a cup of tea." And he squatted down. One of the men poured him out a pannikin. Silently they sipped their tea. Then the Monaro Cocky got up and said:

"Come on."

They all started walking down towards the Flat.

But Joe Summers stepped out of his tent; he had been watching. Joe Summers was a favourite, and a genuine prize-fighter. He came out with Tommy Burns when Burns fought Jack Johnson. Joe fought Jack Blackmore. Afterwards he came up to the field for training where he could be away from the city, and where the unusual work of opal-gouging would help his training. He was a gentleman prize-fighter, very different indeed to the "pug" type. Glad we were for the Monaro Cocky's sake that he now stepped in and took charge. There was an angry glint in Joe's eye; he feared that the Monaro Cocky was in for the hiding of his life.

"Now boys," he said, "this is going to be a fair fight. If any man interferes, he will have to deal with me. Now make a ring."

They did so, the crowd fast gathering. The two combatants stripped to the waist. Young, finely muscled lads; the half-pug confident, a slight sneer upon his tough face. The Monaro Cocky standing quietly determined, his usual smile gone. Joe stepped between them.

"Fight fair, boys, and I'll guarantee no one else interferes. Ready?"

"Yes."

"Then go." And Summers stepped aside.

The fighting man pranced forward crouching, his eyes narrowed, his fists feinting, seeking a shattering opening. The Monaro Cocky faced up stolidly with his fists half raised, staring at the eyes of his antagonist. The fighting man lunged swiftly; there was a crack like the thump of a hammer and the fighting man jerked backward amongst the crowd. We gasped. That fighting man had leaped fair into an awful smack. The Monaro Cocky stood there calmly, only now his face was swiftly flushing.

The fighting man came again with a startled liveliness in his pins, one eye glaring mistily from a distortedly surprised face. He pranced about knocking holes in the air as the Monaro Cocky with an easy swaying movement glided aside from blow after blow. It was a breathless sight, one furious body in swift action, the other fighting with what looked like an effortless gliding movement.

Smack! The half-pug collapsed, sprawling at the feet of the crowd. Wonderful.

The half-pug slowly rose this time; his eyes were shower baths, his face a pained expression. He had stopped two blows that sounded like the kick from a horse. He advanced now more timidly than warily, and his face had changed with the attitude of his body. Behind protecting arms bent almost double he crouched forward unwillingly. The Monaro Cocky suddenly leaped forward.

Crack! The fighting man was on his back, feebly with legs and arms warding off imaginary blows. Joe Summers helped him to his feet while speaking to the Monaro Cocky:

"Leave him alone, Cocky, or you'll kill him. Shake." They shook, the battered pug propped up by Summers's strong arm.

"In future," drawled the Monaro Cocky to the dazed pug, "make sure of your man before you hit him. He may be able to hit back."

Then we all drifted back to our camps for breakfast.

24. KAISER

Kaiser had joined the ranks of the Opal Kings. He had been on opal for twelve months now – interspersed with furious drinking-bouts. There was no hotel on the field in the early years but grog reached there; also it was brewed locally. Some of it was potent enough to burn the leather off a man's boots. It was made down old abandoned shafts which were a warren of connecting drives below. In places on the field now you could climb down an old shaft, crawl through a maze of drives and come up another shaft hundreds of yards away. These catacombs provided ideal hide-outs for the brewing of illicit whisky. Although, in comparison with the big money floating around the field, there was not a great amount of it.

Years of wild life and heavy drinking suddenly took their toll of Kaiser. He was struck down – paralysed.

An awful thing, this once powerful, grim, fightable man, now a savage-eyed cripple, painfully dragging himself along the earth on his elbows. His harsh voice was the same, his fierce eyes the same but – he was paralysed. It was the end of Kaiser, so we thought. They took him into Walgett Hospital. He refused to believe that his case was hopeless.

One night, when the night-nurse was engaged down at the end of the veranda, Kaiser wanted a drink of water. It was some time before the nurse heard him. She brought him a drink.

"You have been a long time coming," he growled.

"Not so long," .she answered. Then stung by the fury in his eyes said:
"Why should I hurry, anyway."

"You think I am going to die," he hissed.

"You know what I think." She took the empty glass from his claw-like hands.

"You are going to be married soon," he said quietly.

"Yes."

"That is a good type of boy of yours, a healthy, strong man."

"Yes, he is healthy and strong."

"You are a good-looking woman, healthy and strong yourself."

"Yes, I am healthy and strong too."

"I will be walking before your first baby is born," he shrieked, "You..."

He abused her fearfully. She fled.

Meanwhile, Kaiser lingered. Refused to die. Refused to give in.

Time sped by. It was announced that a specialist would soon be due in town from Sydney, the hospital was going to get him to have a look at critical cases, among them Kaiser. But every one, and Kaiser too knew it was only a matter of form.

"I'll walk!" swore Kaiser. "I'll walk again!" And the more certain they were that he would never walk, the madder he became.

One day, Maori visited the hospital to see Kaiser. Maori was down and out.

"Can you get a sulky, Maori?" demanded Kaiser.

"I dunno," answered Maori sombrely. "I can try."

"Get one. Get me out of this!" hissed Kaiser. "Get me out. I don't care how, but get me *out!*"

Like conspirators they whispered together by Kaiser's bed. Then Maori rose and with a knowing farewell nod quietly vanished.

It was night. The Walgett streets were dark, kerosene lamps shone dully in houses. Maori was a shadow crouching in the hotel yard. He waited until the last light in the hotel went out, the last roisterers slouched from the yard. Then he crept to the stables. Strange! Not a horse was in the stables. But he knew there was a light sulky in the shed. He had only one idea; he must get a sulky, get it to Kaiser. He manned the sulky shafts, then cautiously pulled it out on to the road.

With bent back hauling steadily on the shafts he trudged on towards the distant hospital. He must have been within a hundred yards of it when he heard a hoarse shout.

"Maori!"

He dragged the sulky across, and the wild eyes of Kaiser glared up at him. He had waited his chance, crawled down the hospital ward, down the path, and down to the road dragging himself on chest and elbows.

"The horse?" gasped Kaiser.

"Can't get one. There don't seem to be a horse in town."

"Hell! and the Ridge is fifty miles away. You can't drag that sulky fifty miles."

"I can drag it a few miles," scowled Maori doggedly. "I'll pick up a horse on the road – somewhere."

"Lift me in."

Maori lifted him in; Kaiser huddled down in the sulky. "Right."

Maori got between the shafts, grasped them, bent his back, plodded steadily down the road. Suddenly:

"Maori!"

"Yeh!"

"To-morrow is coach day?"

"Yeh!"

"Take me back to town then! Hide me in the pub stables. Then when the coach calls for passengers in the morning lift me up on the roof of the coach among the swags, out of everybody's way. The driver will give me a lift, I know him."

"Good idea!" grunted Maori. And he wheeled back towards the black mass of the town.

It was while the passengers were having breakfast in the hotel dining-room that Maori mysteriously sought out the driver. Hoarsely he whispered. They disappeared away down among the stables. They reappeared, carrying Kaiser. There was hardly a soul in the street. They manhandled Kaiser right up to the very roof of the coach, it was one of those big old Cobb and Co. coaches, the roof was only a little below the level of the hotel veranda. They packed him securely among the swags and boxes.

"I'm set," growled Kaiser. And he almost smiled.

After breakfast there followed the bustle always attendant on a departing coach. The driver climbed up to take the ribbons; the passengers climbed aboard; townsfolk on the footpath chatted to departing friends. Just then Dr Cotton and the specialist walked out on to the hotel veranda and leaned over the rail, gazing down on the coach. The specialist had arrived by the morning's train.

"My heavens!" exclaimed Dr Cotton. He was staring straight down into the fierce eyes of Kaiser.

"What is the matter?" inquired the specialist.

"That man packed down there on the roof of the coach," said the doctor softly, "is the paralysed patient. Kaiser," he called softly, "what are you doing there?"

"I'm going back to the Ridge," answered Kaiser defiantly, "and all hell won't stop me."

"Man, you've got the heart of a lion," said the doctor. "I wish I could cure you; I can't. It will be misery for you if you go back to the Ridge. Why not wait patiently here and just see if the specialist can do anything?"

"You know I'm hopeless," sneered Kaiser. "That's what the specialist would say if he examined me. You can all go to hell. I'm going back to the Ridge – I'll walk again."

"Never, Kaiser," said the doctor firmly. "Never if you go back to the Ridge."

"I'll walk again," declared Kaiser madly. "If I stay here you'll carry me to the boneyard. I'm going back to the Ridge. And I'll walk again!"

To a cheery bustle, rumble of wheels, thud of hooves, the coach rattled away. A long, weary fifty miles. Kaiser huddled away up there, watched the dust-cloud rolling up from behind the coach. The murmuring of the passengers down inside, the creaking, the jogging of hooves worked into a fierce, terrible rhythm: "I'll walk – again! I'll walk – again! I'll walk – again!"

And as fiercely Kaiser whispered to the sky: "I'll live! I'll live! ... I'll walk! I'll walk!"

Born May Zinga in Bundaberg in 1894, May was adopted into the Wirth circus family in 1904 and travelled through NSW.

25. THE COUNT

At the Ridge, they put Kaiser into his camp. There he stayed, but now and then he crawled to nearby camps, particularly Nugget Nelson's, on his elbows and chest. Various people gave him food, several threw it to him almost as to a dog, but not by unkindly intention. He was such a difficult man to show sympathy to; it was so hard to do anything for him. He just lay there, glaring up; it seemed easiest to just drop something within reach with a cheery word, then walk casually away.

Always when he crawled to Nelson's camp he would say: "See! I'm living. And I'll walk. I'll walk."

"I believe you will, Kaiser," answered Nugget, and tried to make his kindly eyes show proof. "He's got the heart of a lion; he's all lion," said Nugget to us. "But – he'll never walk. Poor chap."

Unusually heavy rains came that flooded the Narran Lakes and swamped great areas of lowlying country, while the whole bush sang with the croaking of the frogs. A stifling summer followed that was noted for its mosquitoes. In countless billions they had bred in the sodden bush, and in the miles of water-logged shafts. At sunset hordes of mosquito armies made camp life sheer misery. Well before evening we were forced under our mosquito-nets which quickly became a grey mass of mosquitoes struggling to pierce the net to suck the living blood inside.

Sometimes the night became hideous with the howling of a dog being eaten alive; even fowls were killed. Out in the bush, maddened horses galloped the night through or else huddled in restless mobs or pairs thrashing one another with their tails for protection. Groups of men stayed up all night, sweating in panting misery in the stifling smoke of fires which they lit for the tortured horses. The fires were of great logs upon which the firekeepers continuously threw green branches and bushes. Into the thick, suffocating smoke the horses madly crowded.

Kaiser, lying alone in the dark, listened to the ceaseless hum of the mosquito host. The vicious trill, the blood-sucking hate in that angry hum was akin to his own fierce spirit. Long since, he had let them bite his face, his eyelids, his nose, his ears, let them mass on his cheeks until the tortured lips were blood. A night soon came when he could stand it no longer, he rubbed his face on the pillow and squashed them in hundreds. Then clawed the mosquito-net over his head only to the chin, then with his elbows rolled himself on to his back again. Thus he allowed them swarm over his neck until the stinging heat became unbearable. A night soon came when he pulled the mosquito-net down to his neck and waited, his eyes rolling madly in the dark, his naked body a mass of ravening mosquitoes. He could just feel a few tiny stings at the base of his neck and a burning on the shoulder-tips, that was all – but it was enough. The remainder of his body was a log, a pulsing mass of biting mosquitoes. But – he could not feel them.

Night after night he lay thus, glaring into a hum of sound. Then a night came when his teeth grinned to the roll of his. eyes ... Yes, by God! He felt his shoulders slowly beginning to burn. He waited, open-mouthed, mad-eyed. Yes! He shrieked – once. He was sure. He could *feel* them! He dug his elbows into the bunk, laughing in a terrible, chuckling silence. He could feel the mosquitoes biting nearly half-way down his arms, nearly half-way down his chest.

In a month, he could slowly use his arms. He was forced to lower the mosquito-net more than half-way down his chest. He could just feel them biting then, the skin on his stomach just faintly tingling, beginning to live again under the bites. He breathed slowly, taking in great breaths, distending his stomach so that the taut skin stretched the better to take the million poisonous lances thrusting into it. His stomach was a living plaster of grey mosquitoes, standing on their snouts as they pressed their lances deep – deeper.

Another month dragged by. He had been forced to lower the net down over his hips, his stomach was raw meat. But, by God, it lived!

And now, by day, he crawled about his camp exercising his strengthening arms, his straightening legs. Then, while the men were at work he crawled on straight arms now and upper body ever so much easier, but still dragging his legs, crawled out among the camps. And crawled back with a couple of sticks and some string. He fashioned two excellent crutches, chuckling over his handiwork in a delight he could never have felt had he been handling the most beautiful opals.

That terrible summer was almost gone, cool weather was coming, the mosquitoes, thank heaven, were dying out. Nugget Nelson was lying in his camp one evening absorbed in a book; he had always been a great reader. He looked up startled by a wild peal of laughter. On crutches, Kaiser collapsed into the camp shouting:

"I walk! I walk!" Glaring up at Nelson, he laughed hysterically, and kept on laughing.

"My God!" whispered Nelson.

He slipped out of the bunk to help Kaiser rise. But Kaiser waved his arms at him and howled: "I walk!" In a month he had discarded the crutches. Two weeks later he commenced sinking a shaft.

Kaiser completely recovered. It was a miracle. We could only imagine that the terrible irritation of those ceaseless, countless biting lancets had tortured life again into the living mummy; had again started the circulation of the blood.

I'm writing these reminiscences half a lifetime later but cannot resist the temptation to skip years and tell you the end of Kaiser. He was always the same fierce Kaiser. He made thousands and went through it all again. After the war, I met by chance an old Lightning Ridge man, in Queen Street, Brisbane. We talked of old times:

"By the way, ever hear anything of Kaiser?" I asked.

"He was my trench mate in Flanders!" he answered.

Years later, he turned up at Dirranbandi, in Queensland, driving a drover's cart. The same implacable Kaiser, with still harsher lines chiselled down his face. He sought out Mick Cullen and Nugget Nelson who had taken up land near Dirranbandi. They have turned that land into Tootherang station, and live there to-day.

Throughout the years, Kaiser, when travelling through Dirranbandi, always sought out Mick and Nugget. And always his plaint was that a man was a fool wasting his time in the sheep country; he ought to be away out in the Territory looking for gold. One day he went. In his last talk with Nugget Nelson he said:

"They call me Charles Anderson. But no one will ever know me, even though there's a mark on my body I'd always be known by. I'll die too far away. When I do go I hope I die far out where the dingoes will pick my bones before they find me."

Several years ago the Queensland newspapers printed a little paragraph: "An unknown prospector, believed to go by the name of Charles Anderson, has been found dead on the lonely track between Brock's Creek and Birdum, Northern Territory. The dingoes had partly eaten the body."

Tom and I were still battling, digging for opal. Around old Matt's campfire we wondered at times whether the Count too was one of the Legion, the Legion of the Damned.

He was a real Count, a likeable man, a great mate of John Landers. A remittance man, the majority of the gougers only knew him as an Austrian Count who had "done something or other" and was now an opal-gouger. But at night when foreign mail came for him he confided deeply in old John, and one restless night told all – or nearly all – the story:

He was Count Hoyst, and the story was of the greatest tragedy in the tragic lives of the Emperor Francis Joseph of Austria and the Empress Elizabeth. This was the murder or suicide of the Crown Prince Rudolf of Austria and his beautiful sweetheart, the Baroness Marie Vetsera.

"It was a terrible night," said the Count sombrely, "and yet a night of joyousness, of laughter and of love. The Crown Prince had played truant from the Emperor and the Court, and was giving a wild party to his host of brilliant friends at his hunting-lodge at Mayerling. Deep in sombre woods that beautiful Lodge was hidden from the critical eyes of the world.

"It was a scene in Fairyland, that ballroom and dining-hall. Beautiful women flashing with jewels, tall officers in gorgeous uniform. How the wine flowed! Music and laughter, song and jest. Brilliant light that melted on the lovely shoulders of the women. A dream of loveliness was the young Baroness Marie Vetsera. I can see her eyes now, eyes that a man would risk his all for. She was sitting by the side of the prince; he toasted her again and again. But the toast was bitter-sweet. Among the guests was a young aristocrat madly in love with the girl – and the Prince knew it. He would put his arms around the baroness, drink to her with his eyes and the glass, and kiss her. At last the young lover rose from the table and hurried from the room – mad with jealousy."

The Count stared for a long while, saying nothing. Then he spoke huskily:

"Next morning the Prince's valet Loschek went to call the Prince. He tapped at his door. No answer. He tapped again: tapped urgently, thinking the Prince asleep. He *was* asleep. My own brother, Count Hoyst happened to come along. They tried to wake the Prince. In vain.

"'Step back,' ordered my brother, 'then spring forward with me and crash our shoulders against the door.'

"They did so, and burst into the room.

"The Crown Prince was there, half sitting. A revolver had fallen from his hand. Marie Vetsera was lying there, her beautiful hair all loose. She clutched a flower in her hand. Both were shot through the head."

"And the verdict?" asked old John softly.

"Suicide," murmured the Count.

The Count told the story more than once until old John asked:

"But what became of the young lover?"

"He was never seen again," answered the Count evasively.

[Before letting this particular story go into print I took the trouble to read up the history of the then reigning House of Austria. The story was there, just as the Count had told it. But the Count who had burst open the door spelled his name Count Hoyos. We called our Count just "the Count", or "Count Hoyst". Further, there was no mention in the accepted story of the royal tragedy of the young lover.]

Tom and I toiled in the bowels of the earth still seeking the fiery stone. It is wonderful, that such a dream of beauty, carrying the concentrated loveliness of the sky and sun and stars should be imprisoned in utter darkness here deep within the earth. Everything must have been created from some vast Mind, and every creation was a perfect thought: the perfection of the seed in the flower, the sheen on a rifle-bird's wing, the marvellous eyes of the eagle. The formation of a stone, the chemical excellence of a gas, the metal in the grain of tin – these were perfect creations. So, too, was opal, formed many millions of years before man came upon the earth. For many feet under solid rock here, we occasionally dug out opalized seashells.

One smoke-oh Tom leaned back and remarked: "Does it ever occur to you what insects we are?"

"Not particularly. What bug is biting you now?"

"Oh, just some old sea memory of a hundred million million years ago. Here we are under a hundred-foot depth of solid sandstone, we've had to use gelignite to blast our way down through this rock. A few hundreds of millions of years ago this rock was sand on the bottom of the sea. You've got the proof in your hand – you've just dug out that opalized fish!"

"That's so. But what about it?"

"Oh nothing. But there can't be such a thing as time."

"Bullswool! It's a jolly long time since this fish was swimming in the sea."

"Fathead! What if it is! It'll still be a jolly long time, millions of years after your bones have dissolved into the finest dust. Time won't count then any more than it counted when this fish was in the flesh."

"I wish the dashed thing was pure opal instead of being an opalized fish half stone and potch. If it was only pure opal I'd jolly soon sell it, then quick and lively have a jolly good time."

Tom was disgusted.

It was fascinating, all the same. The world a vast sea. The sea in places dries up. The sand is turned into solid rock. Far below that rock wee stones of nature's glass form and in those little stones is imprisoned light and colour more beautiful by far than the colours of any living thing under the sun.

Who made the opal? Who put the colours in the bird? in the butterfly? in the flower? in the eyes of a child?

Deep down below, Bill Younger dug out opalized wood. So that, after a sea, a forest must have grown here too. Gougers one day dug out the opalized fossil of a bird. So that birds must have flown in the forest. Then Mick Cullen and George Bailey dug out of Tingha's claim the opalized bones of a little animal that was nearly all teeth, beautifully coloured opalized teeth.

I was always hoping that some day we might dig out the opalized bones of a man. But we didn't. Sea, forest, birds and animals had vanished, pressed out of existence by the steady formation of a hundred feet of solid rock. Creation thus gradually preparing the way for man to come upon the earth.

Tom may have been right. What did time now mean to the countless millions of creatures that had lived where now we dug deep under the solid rock.

Opal cutting, Lightning Ridge.

26. THE OPAL-BUYERS

One midday Mick Cullen came hurrying to the claim: "Big Ben has gone bush," he said. "The horrors! They're forming search parties." For once he wasn't joking. The weather was hot, a crazed man could easily perish. Search parties formed swiftly.

Ted Moody was already tracking. He was almost as good as a wild aboriginal. Tall and slim, with wonderful eyes that instantly could see a story in an overturned leaf, a broken twig, a misplaced pebble, a crushed blade of grass, Moody hung to the tracks with the tenacity of a bloodhound. But the demented man was possessed by an imp's cunning. He had circled in among the camps, among the dumps; by every trick his mind suggested he sought to hide his tracks while his crazed brain whispered:

"They're after you!"

Tom and Mick Cullen, two of the author's fellow gougers at Lightning Ridge. Mick eventually took up land in Queensland and started a sheep station.

Moody cut the tracks. Very slowly, very patiently, he followed them. Miles through the bush they led on to a strip of hard country. And there the tracks vanished. Moody presently solved it. The crazed man, maddened into some devilish cunning – had taken off his boots. But he forgot his desperate hurry. His toes displaced a pebble here, a twig there, trod down a grass tuft here, rolled over a little pebble there. Steadily, slowly, Moody tracked him miles away out into the bush, then the tracks vanished again by a wire fence. Utterly vanished. Presently, Moody solved it. The cunning man had grasped the top wire and walked along the bottom wire with his bare feet. He had gone far along the fence in that way. It took Moody a long time to find where he had left the fence and taken to the bush again.

Next day, tracking steadily, all hands kept a sharp look out. Big Ben was a powerful man; he was crazed; there might be a struggle when they found him.

Meanwhile, far ahead in the bush, the crazed man was crouching, staring into the night. The night was a thing alive; big shadows with groping tentacles reached out for him. Some shadows had eyes that grew as large as water-lily leaves and balefully glowed with a pale green phosphorescence. The mosquitoes were all coming to him urgently humming: "They're coming! They're coming!" He sprang up and shrieked and ran through the bush. His brain was on fire, his tongue was brimstone, his feet were cut and bleeding, his heart was thumping madly. They were "Coming! Coming! Coming!"

Daylight came. There was a peewee up in a tree. Suddenly it flapped its wings and shrilled: "They've got you! They've got you!" He shrieked, and doubled back into the bush.

All that morning he crouched, and ran, and crouched, and ran. He was spelling under a bush when a wallaby bounded by and the thump of its tail urged him: "Run! Run! Run!"

He ran. Exhausted, he dropped under a tree, burying his blistered lips to the earth. An old black crow flopped heavily in a branch overhead and said: "Kark! Look out! ... Kark! Look out!" He staggered to his feet and ran blindly, bumping against the trees.

He fell down at last. But the old crow was his friend; it looked after him. It followed him again and settled in a tree above him and warned: "Kark! Look out! ... Kark! Look out!" Again he ran and the crow still followed, warning him they were coming. As the afternoon wore on, other crows joined the old crow and karked him warning every time he fell.

Darkness came again, just when he was done. The coolness revived him. But he couldn't run any more, he just stumbled along, then fell in the darkness. Daylight came. He staggered to his feet as a soldier bird suddenly screeched out: "Look out! Run!" He ran, ran in mad terror as the wild bull chased him. Reaching a forked tree in time, he leaped up with the bull's horns clashing at his feet. When he was sure the bull was gone he climbed down, and staggered forward. He was sure it was the wild bull that had chased him the night before.

Suddenly, he came to the edge of the bush and stared unbelievingly, trying to realize what instinct was urging. But he daren't leave the timber. He was staring at the Narran Lakes. A gleaming expanse of water and reeds, with wild ducks upon the water.

Suddenly, two wild pigs rushed past saying: "Whoof! Whoof!" In choking terror he staggered back amongst the timber.

But instinct drew him back, a terrible, torturing instinct. He crept back, then began crawling towards the lake on hands and knees, his bloodshot eyes terrible to see, his parched mouth wide open, gasping; his leathern lips could not help him breathe.

At last he reached the mud, crawled away out to the water, and tried to drink; he could not, he crawled out into the shallows and lay in the water. He lay a long, long time, the water gradually soaking into his body. He kept sinking his mouth into the water and at last a little of it began to soak down.

He lay there for some hours and the water brought him to. After nightfall he realized what had happened. He felt terribly ashamed. Crawling back to the edge of the lake he thought a long time, then went to sleep. Next morning he drank steadily; thought out his bearings; then doggedly started to walk back to the Ridge. He realized he must have travelled, circling and all, at least seventy miles. He blessed his great strength, but he was a pitiably weak man now. When the sun heated the earth he suffered agonies from his torn feet.

Steadily he walked, forcing himself to keep walking; seeking where possible the shady way, carefully nursing his strength.

At midday he saw a man coming surely towards him. He stared hard at the man, the man was staring at him. Suddenly fear gripped him. As they met, he whispered urgently: "Are you real – or imaginary?"

On the field there arrived a man, Sparks we called him. He was a wildly enthusiastic, cheery young optimist. He brought the first motor-bike to the field. It startled the bush with explosions; poisoned it with smells and puffs of horrid ·smoke. "He'll blow himself to pieces," declared Tom, "and others with him." It was an awful machine, that first motor-bike; a stubborn, "I won't go", fair cow of a machine.

Strange, looking back now. Smooth-running cars are a commonplace of the day; fleets of aeroplanes roar through the sky. Sparks and his spluttering, roaring, stinking, back-firing, exploding, buckjumping, whistling, smoking, infernal machine seem far, far away.

Who in that not-so-long-ago imagined fleets of aeroplanes, moving pictures and wireless? What on earth will the world be like, what inventions will there be, in another hundred years from now?

Tom and Mick went all literary again, urged me to scribble paragraphs for the *Bulletin,* for the Abo. Column. They simply refused to notice ridicule or laughter. Night after night lying there smoking in bunk Tom would return to the attack.

"You've got lots of material," he persisted; "little bush incidents of interest about bush characters you've met, or animals or birds or snakes."

"Not interested," I'd yawn. Then Tom each week brought the *Bulletin* to camp. We grew to look forward to the Abo. Column. At old Matt's camp we'd yarn over the pars by night. Those quaint paragraphs were written by bushmen about the bush we loved.

"You can write paragraphs just as interesting, laddie," Tom persisted.

"Give it a go," said Mick.

They won out at last. But the first paragraphs were not accepted. Surprised, I tried again with like result – a disdainful silence.

"They don't know good stuff when they see it," I declared scornfully.

"You can't make it!" Tom chuckled and spat straight out the tent door.

I sat up and scribbled a dozen pars before dousing the glim.

Those pars weren't accepted. I became interested; studied the Abo. Column. Then wrote again. Later, Tom came hurrying up from the mail, the "red rag" under his arm. He opened it at the Abo. Column and his none too clean finger pointed to the pen-name "Gouger".

It was quite a bit of a thrill; almost a sort of triumph.

"Now keep it going," nodded Tom.

I didn't. Patiently he renewed the attack. Eventually, at odd moments, I wrote a par or two. It grew to be like a game, almost like opal-gouging. You sank a shaft, hoping, expecting opal would be on the bottom. It wasn't. You opened the Abo. Column, hoping, expecting, the name of Gouger would be there. It wasn't.

Disappointed, you sank another shaft. Disappointed, you wrote another par.

As the months grew into a couple of years, pars appeared occasionally. It was to be a few years yet before practically every issue of the *Bulletin* held different pars of mine under different pen-names. And very handy the returns for those pars were to prove. At times I've ridden in, completely broke, to some tiny bush-mail place to find a *Bulletin* cheque for twelve-months pars awaiting me. And I'd be on my feet again.

But in the Ridge years, Tom had to drive – drive – drive. Once he got me enthusiastic enough to scribble out an article, and a short story. The rejection of both appeared in two short, sharp, scathing sentences in the "Answers to Correspondents" column. For sheer blood-curdling vindictiveness, for scathing ridicule, those two refusals were gems.

I gasped; then laughed and mentally vowed to make the *Bulletin* eat its words; a time would come when this wretched red rag would print my yarns, any amount of them.

The time did come, but not for years.

Meanwhile, Jack Scott and Harry McCullum had bottomed on a patch worth £13,000. They got £100 for the chips they knocked off the first stone. Imagine the marvel of that stone, shattered by a pick, the fragments sold for a hundred golden sovereigns! McCullum had been looking for a mate for a long time. He tried to persuade Nugget Nelson to go in with him, but Nugget had had a run of bad luck and wanted to go away shearing. None of us liked going mates with a man unless we could pay our own way. McCullum asked Jack Scott to go in with him. Jack wouldn't say yes for a long time, for he knew that a man who had half promised to help McCullum sink the proposed hole was away.

"He has gone," insisted McCullum. "Gone bush and probably working somewhere. If you don't help me sink this hole I must get another mate."

Still Jack hesitated, because of that other man ... "Well, if he doesn't come back after next week-end, then I'll go in mates with you," at last agreed Scott.

They sank the hole, and bottomed on a small fortune.

After Nugget returned from the shearing, Jack Scott called him aside one evening. He opened a shabby old tobacco tin and the beauty of the world flashed to the stars. Nugget held his breath; it was a sight that would make any man gaze reverently.

"What do you think they're worth?" murmured Scott at last.

"I don't know," answered Nugget softly. "I can't imagine what they look like in daylight, but at night they look worth £100 a stone."

There were eleven stones. They realized £1200.

To-day, each stone would be worth £1000.

Tingha's claim was famous; stars of beauty were coming from there. Old Matt and Jimmy Stedman and Paddy Kelly faced a rich parcel from up the Gully. Parcel after parcel was unearthed and faced; hundreds of pounds worth – thousands of pounds worth – of loveliness in parcel after parcel. The gems among them were spoken of with bated breath. Now hundreds of men were on opal, Murphy was buying it by the thousands of pounds worth. Solomons was travelling backwards and forwards from the field to the cities, buying and selling, as fast as he could travel. He had come to White Cliffs hawking little glass images for a living. One day a gouger was examining one and clumsily broke its arm off.

"Oh!" exclaimed Solomons, "You've broken the arm off my little doll!"

"Well, poke his eye out and sell him as Nelson," suggested the gouger.

The dapper Bottomley, stepping out like a figure hopping from a bandbox, was now familiar with the great cities of the world. The Dominick brothers, Abotomey, George Cowan, Andy Sharp were buying well.

Ernie Sherman was getting together a magnificent parcel of black gems worthy of a king's ransom, to take to London, and then to India to the great Delhi Durbar. Percy Marks was investing heavily.

Energetic little "Berlini" again and again startled the field with his gambles in rich parcels of stones. A livewire Berlini, a cheery, well-liked little bundle of energy whose delight was a dazzling gamble. He bought many a parcel in the rough that faced into beautiful gems, but he suffered many a disappointment. He gambled periodically on racehorses in the distant cities. He was a born gambler; I believe he would have gambled his life with a smile. But it was shrewd, speculative gambling. "Life was short; full of ups and downs. It rested upon a man's intuition, sense, and gameness as to whether he experienced more ups than downs."

That was Berlini's philosophy. Again and again at the Ridge he went broke, but very soon he would be up on top of the world again, buying parcels for thousands of pounds. He liked to tell of the night his first child was born. It was at White Cliffs. Berlini was worth just seven and sixpence. He hurried to a poker-school, and won £2. Six months later, he was the biggest buyer on the Cliffs. To accomplish that meant phenomenal luck or phenomenal judgment, whichever you like.

Smiler Lawless told a tale of Berlini at the Cliffs: "I sold Berlini a parcel for £1500. He hadn't the cash at the time, so sent the parcel to the city and waited for returns. One day we were in a two-up school together; the school soon grew pretty big; there was plenty of money flying about. I got the kip for the toss and said to Berlini:

"'You owe me £1500 for that parcel of opal, Berlini. I'll toss you –

double or quits.'"

"'Right!' said Berlini.

So we tossed, £3000 or nothing. Berlini won the toss."

All hands liked Berlini. He fought for big money and when he got it he paid for big parcels as if there was never going to be any settling. And when he went down, he went down with a smile; we knew Berlini would come up again. He always did. Only one thing ever got Berlini down and out – death.

He sleeps at the Ridge. No epitaph. But the little mound is bright with potch and colour. He lived among opals. He sleeps among them.

Three Mile Rush, Lightning Ridge 1909.

27. THE OLD MAN

An old, old man used to trudge wearily past our camp every morning while we sat at breakfast, disappearing among the bushes up the Hill. Every evening with bowed back he would come shuffling down past the camp as we sat at the evening meal. Evidently he camped down the Flat, somewhere near Andy the Swede's bag camp. The monotony of it, the thin, bowed-shouldered helplessness of the old man, the hopelessness of his little white face got on my nerves.

"I wish that old fellow wouldn't go mooning past our camp every day like this," I remarked one evening. "He takes away my appetite."

"Mine, too," mumbled Tom. "I wish old Dad would keep to the Flat. There're any amount of shallow holes he could potter about in there."

"Why does he climb up to the Hill?"

"Oh, he's got a bee in his bonnet. There's an abandoned hole up there; he thinks there's opal in it."

"I didn't know there was a hole up there."

"Neither did I until he told me. Some prospectors put it down. Seventy feet deep too. Sank it to the opal dirt, then abandoned it."

"They must have been super optimists to imagine there was opal up there."

"Yes. There's no opal within a mile of it."

"But what does the old chap do up there?"

"Sits on the shaft dump, hoping someone will come along and help him sink the shaft to the second level."

"Good heavens! Even if the shaft wasn't bottomed no one would dream of sinking there. And if they did they couldn't take the old man as mate. He couldn't wind a bucket, let alone swing a pick. Why does he sit there?"

"Because he can do nothing else. He is sure there is opal away below in the second level, and he just sits there hoping for someone to help him. He has asked nearly every one. But they take it as a joke, and try to get him to keep down on the Flat among the shallow holes."

"It's a pity he wouldn't. What a mad idea! Opal up on the Hill! And a second level! Why there mightn't be a second level for a thousand feet – ten thousand feet ... Has he asked *you* to help him sink the shaft?"

"Yes," answered Tom shamefacedly. "It's a three man job. He wanted both of us."

"Well, I'm blessed!"

Day after day, week after week, the old man still toiled up the Hill. One evening when he was trudging past the camp I called him over to the fire.

"You believe there's opal up on the Hill, Dad?"

"Yes!" and his old eyes shone eagerly.

"Why?"

"I know!" he quavered. "I know."

"How do you know?"

"I just know."

"Nonsense. No man can know there is opal anywhere until he actually bottoms or drives on it."

"I know," he whispered with tense earnestness. "I know there is opal there – down on the bottom level."

"Why, you don't even know that there *is* a second level."

"I know!"

"Now look here, Dad. You know that a second level has only just been discovered on the field. It's rare, or appears to be rare in the first place. And, you know that after going down through the opal dirt, eventually you strike sandstone again. It may go down hundreds of feet. It may go only twenty – if there is a second level. And you know that that shaft up on the Hill has only been sunk just to the opal dirt. Now, how can you see straight down through that stone and tell us there is another layer of opal dirt underneath?"

"I know!" he whispered. He was almost crying.

"Better give it up, Dad; It's useless." I couldn't look at his face so stirred the fire instead. "You'd best try the shallow holes in the Flat, Dad. You can work there easily, and no trouble."

"Will you help me sink the hole?" he almost whispered.

"Heavens! No, Dad."

He turned dejectedly away and almost crawled back down the track.

Down at Gadsby's that evening we saw some wonderful parcels, beautiful stones that seemed laughing to the lamplight in a joyous abandon of dancing colours. We felt a bit miserable. No matter how hard we worked, no matter where we tried, we couldn't find opal. And all through this evening we were gazing on thousands of pounds worth.

"Come on," said Tom at last, "I'm going back to camp. I want every one of those stones; each is more beautiful than the other. Come on back to camp where I can be properly miserable. Ugh! Dark as the pit. Lightning, and it has started to rain. That's fine! A big storm coming."

One-eyed Fred and Big Ginger were also staring out into the black night. Fred had his right eye out, Big Ginger the left: "Come on," growled

Big Ginger, "let's go. We'll have *one* eye in the right direction anyway." They clasped arms, and stepped out.

We followed them until they turned down a side-track, then carried on up towards the Hill: "We're so lucky," declared Tom with bent head against the rain, "that we'll probably fall down a shaft with opal in it and break our necks." Then a blinding flash illuminated a terrifying sight, a tent blown to tatters, a big tree falling to crash across the ridge-pole.

"God!" exclaimed Tom in the darkness.

We hurried through the sopping bushes to the flattened tent, called out and received the answer we expected. The tent was squashed, the bunk flattened to the ground. We felt along the clammy tree-trunk, felt for head and legs.

"I know!" exclaimed Tom. "Yesterday he bottomed on potch and colour. He might have struck it! He might be working night-shift, afraid of the Ratters!"

We groped our way across to the shaft. Brown water was pouring down one corner of it, gurgling and splashing away below. "He can't be there," said Tom.

"Hullo!" came a muffled shout. We peered down but could see nothing; only hear the water pouring down.

"Down below!" yelled Tom.

"Thank heaven!" came a shout. "Lower me a rope quick! The steps have all crumbled away with the water! I can't climb up!" We lowered a rope and quickly hauled him up. He looked like a drowned rat, but his face was very thankful: "I had an awful fright," he said huskily. "The water was up over my belt; thought I was going to drown like a rat in a trap. Thank God you heard me shouting."

We had heard no shouting. But for that lightning flash and that tree falling across the tent, we would now have been in camp, under the blankets.

He sat very quietly when we took him to our camp and boiled him a drink of tea. We turned in and smoked awhile, then Tom doused the glim. "Ah, well," he said to the darkness, "better to bottom a duffer than strike opal and be drowned."

Day by day, while driving underground down the Gully the thought of that lonely old man sitting up there among the bushes on the Hill began to almost haunt me. Tom's pick would be rhythmically thumping, softly echoed by the thumping on his heart. Sound carries strangely like that, when working down below. Tom would be sitting back on his legs, his body gleaming with sweat, his lined, clear-cut features outlined in the candle-flame as he worked steadily at the face. Squatting back on our heels, we'd shovel the dirt over our shoulders where it thumped back behind us into the shaft bottom. Then when the face was clear of mullock, we'd sit back, roll a cigarette, and yarn.

"I'm a fool," I ventured one morning.

"I'd hate to say 'I told you so,'" murmured Tom.

"Look here, old Dad has got right on my nerves. What if we rig a windlass on that hole and sink the shaft to the second level. Then that spectre will cease haunting us."

"I'm on."

"Then you're as big a fool as I am. I believe you've dashed well wanted to sink it all along."

"No I haven't; you know I don't love work. But that picture of misery has got on my nerves, too."

I took up the pick, and swung again at the face; "Right-oh then. You tell him this evening."

The old man was so excited he wanted to walk around our camp all night. We had to almost force him away.

In shamefaced fashion we carried our tools to camp, then when all hands had gone to work carried them up the Hill. We didn't want any one to know we were such fools as to sink a shaft up there. When we started working only a wanderer might drop across us, for box-trees and buddha and leopard wood and bushes grew thickly on the Hill.

We rigged the windlass, then Tom lowered me down the hole. I examined the opal dirt curiously – it really was opal dirt. Few men would have believed that there was opal dirt under the Hill.

We set into the job. Punched her down through seven feet of opal dirt, then struck sandstone. We went through twenty-five feet of this. The old man just sat up on top all day long, pathetically anxious to do something. Impossible. Tom could wind the windlass twelve times to the old man's one painful turn at the handle. As for using the pick down below – we let him try once, just to convince him.

We punched her down through the sandstone then broke through a band on to a second level. It was a great surprise that a second level was really there. The fact only aggravated the disappointment: the shaft was an utter duffer. I climbed up the shaft.

"It's a duffer," we explained to the old man. "We will take the windlass off to-morrow. But you'll be satisfied now."

He said never a word. Just sat bowed there, his trembling hands in the duffer dirt we had sent up the shaft.

Tom and I took the tools away and began again in the Gully. We had been fools; but it was a real relief to have the old man off our minds.

A week later he slowly climbed the Hill again. I looked at Tom as the old man trudged slowly past the camp. That evening, he trudged slowly back again. He wasn't satisfied. Still thought there was opal up there. Still sat on the dump, day by day, alone amongst the bushes.

Charlie Nettleton, a miner from White Cliffs, saw the potential in black opal in 1903, in a partnership with Jack Murray.

28. ON OPAL

A month later I made up my mind. This would be the last. Almost cheerfully I said to Tom: "I'll tell you what we'll do. We'll drive the first level of the old man's shaft, and throw the dirt down into the second level. Thus we won't have to haul up the dirt. In a few days we'll be mullocked right up. Then we can pull out. The shaft will have been sunk to the second level, the first level will have been driven say fifteen feet anyway. A thorough test."

"Right," agreed Tom.

In three days we'd driven the shaft. It was easy. Just driving, then throwing the mullock over the shoulder, where it fell down the shaft. In three days it was mullocked up to the drive.

Tom had climbed up the shaft, and hauled up his tools. I was still gouging, hampered now by the quickly accumulating mullock. Very soon, it would be impossible to work any more. Tom would be trying to cheer the old man on top.

Goodness knows what made me delay, and delay there lying on my side, gouging, gouging farther in until I could barely move for the mullock. Sand from the roof now and again fell into my eyes and guttered the dim candle-flame. Lying crouched there side on, the roof pressing me down, I held the candle to the face. Up near the roof, through the greyish-white opal dirt several little rusty looking bands had come in, like a drip of red dust. I chipped the bands with the chisel point, they felt gritty. This was faintly interesting. The air was terribly hot. There was hardly room to breathe. Jamming the spider into the wall I lay there gouging steadily in, scraping the fallen earth away by hand, just able now to gouge in under the roof.

Click! The old heart missed a beat. I stared, then slowly held the candle to the face.

A bead of dancing orange was glowing from the face!

Thump! Thump! Thump! Thump! The old heart sounded quite loud in the deathly silence down there; it seemed suddenly to have swollen and to pump slowly and hurtfully. I lay back for a long time, staring at the roof within six inches of my nose.

Opal! We were on opal! Oh, the delight of being on opal! I gouged out four little stones. Impossible to work any more until the shaft was cleaned out. Gasping, on scraping elbows, inch by inch I crawled backwards from the face, smiling in a silly happiness.

It must be jolly late. Yes! There was Tom's face away up there outlined by a sky from which stars twinkled. It was well after sundown. Tom had been peering down there a long time, too incredulous to allow himself to imagine things.

That was a smiling, joyous climb, step by step up that shaft. They would be wondering – wondering. An opal would be coming up the shaft. Tom didn't shout, just waited for the climber to come right up and crawl out on to the dump. Twilight had come. Tom was staring; the old man was crouched away down the dump.

"Dad," I called, "come up here!" As he came labouring up I looked at Tom, waved my hands:

"Opal!" I said.

"You don't say!" Tom whispered. I showed them the stones. They could only stare and stare and stare. Old Dad couldn't hold the wee stones, his hands were trembling so. His teeth began to chatter.

"Come on! It will be dark before we get back to camp." Then I turned on the old man. "Don't you dare breathe a word of this, Dad," I said harshly, "or the Ratters will rat the claim!"

He shook his head vigorously.

"If we don't say a word, we are perfectly safe. No one would believe we had struck opal up on the Hill."

Dad was fearfully excited. He tried to eat at our camp; he couldn't. Then he wanted to sit by us with his skinny old arms trembling over his knees. We feared a visitor might come any moment and jump to conclusions after one look at Dad. We drove him away to his own camp at last. Tom and I had got such a surprise we hardly slept. To think we were on opal! And up on the Hill!

Next morning well before daylight we heard the fire blazing, shuffling feet; someone putting the billy on. It was the old man.

"You shouldn't have done it, Dad," remonstrated Tom. "You'll make every one suspicious." Old Dad nearly cried. We went to work as usual. I climbed down; we were feverishly eager to get to work and clean out the shaft.

Heavens, it was mullocked up all right. It was a wonder how I'd worked and breathed so far in under there in the drive. Down came the bucket, I stood it in the corner atop of the pickhandle and began shovelling in the mullock. At the third shovelful I stared – gleam of orange and green lay on top of the mullock on the shovel – the fragment of a lovely stone.

I went cold all over, could have almost cried. Lit the candle and crawled into the drive, staring at the face.

Ratted!

They'd come down in the night and worked like fiends. Two men up above continuously on the windlass, one below shovelling the mullock into the buckets, one gouging into the face; another man behind him throwing back the mullock so that the driver could drive, drive, drive.

The face was gouged right in to the roof, they'd driven eight feet in a night. Broken chips of opal gleamed in the face, chips flashed upon the mullock wherever the candle glow reached. Dejectedly I crawled back to the shaft.

"What's the matter?" shouted Tom.

"Ratted!"

"Wh–what?"

"We've been ratted!"

I climbed up. Old Dad stood bowed over the windlass, staring across among the trees.

"Dad," I accused angrily, "you let it out last night! You told every one we'd struck opal – and the Ratters came and ratted us!" His lip quivered. "Did you, Dad?" demanded Tom. "I–couldn't help it," he whispered.

I sat on the dump and rolled a cigarette. Among the trees four men were climbing up the Hill, tools on their shoulders.

"Dad did you peg out this claim?" ... "Yes."

"Did you peg the shaft in the centre of the claim?" ... "Yes."

"Sure now?" ... "Yes."

"Right."

We should have looked for ourselves. Trees and bushes were fairly thick, we couldn't see the pegs from the dump, we should not have trusted the weak memory of this old man. If the shaft was in the centre of the claim then opal was in the centre and the ground all around to the four boundaries would probably be opal-bearing. But if the shaft was towards any one side of the claim then all our unworked ground might prove duffer ground. We should have looked for ourselves.

The new-comers came to the dump edge, threw down their tools: "Well, you look pretty glum for Opal Kings," smiled one.

"How do you mean?"

"We heard you were on opal."

"How did you come to hear?"

"Oh, there's a rumour spreading all across the Flat and up the Gully."

"H'm. Well we might have been on opal."

"What do you mean?"

"Ratted!"

"What! ... That's stiff luck!"

We smoked on. They meant their sympathy. Other gougers came

along. Old Dad had done his work well. Then the news spread that we'd not only struck opal up on the Hill – but had been ratted too!

That started work on the Hill in earnest.

We cleaned out the shaft, picked away the "toe" that the ratters had left, squared everything up, then next day commenced driving. Other men began sinking among the trees around us, in a few weeks shafts were to bottom on opal; the boom of "the Hill" was to begin. But it was bittersweet to us.

We drove the shaft by night, but the opal very quickly cut out. The Ratters had got the patch, besides smashing hundreds of pounds worth of gems in the robbing of it. We only got the tail end of the patch, just the few stones they could not gouge out before daylight came. By mutual consent I took charge of the stones, and began facing them on Jeff's machine in his hut, now owned by little old Tommy Turley. This was the opal game we'd so often dreamed of. First the finding; then the digging of the stones; then the snipping, the classing of the stones in firsts, seconds, thirds, and finally the potch and colour.

And now came the facing. Picking up a snipped stone, examining carefully the splashes of colour wherever it showed, calculating whether it would be a black or light stone, judging by the grey or light or black potch, trying to tell which way the colour bar ran, endeavouring to make certain of the right surface to face. Then holding it to the carborundum wheel and watching the dirt then potch disappearing under the revolving wheel. Then dipping it in water and smiling with delight as green and orange flashed up, then a flush of valued red, still obscured in great part by potch. A little more grinding, another examination. Brilliant colours now, flashing with fire too and – showing right to the edges of the stone. Almost a certainty now that it *was* going to turn out a stone and not a sandshot blob, a worthless disappointment.

A more careful grinding, a touch of the wheel here to bring out the green, a touch there for the orange, a gentle turn or two here for the red which is now developing into a "flame". The dipping into the water, then a long drawn breath. A gem! Surely. And harlequin pattern! With a depth of colour of infinite variety in soft deep tones quivering and rippling in star glints of orange and flame. Would it face all over thus? Would the last of the scum grind away and leave in perfect beauty this pulsing dream of loveliness?

I'd put on another wheel, a finer one. Only a couple of turns every now and then, numerous examinations, each turn of the wheel licking off more potch, or the hard scum which blinded the colours as a cataract dims the eye.

An anxious hour, grinding deeper into the colour now blazing within the stone. But one turn too much and that colour might vanish. I'd grind then with coarse emery paper, then with medium, then with fine. And a man's heart would be in his mouth for except one wee spot in the dead centre, the stone was a glowing beauty of colour in perfect harlequin. A gem, if only —!

At the very centre, the size of a pupil in an eye, clung a smoky scum spot, growing like a poisonous film right into the very colour. And the colour there was not the thickness of a finger-nail, it fined down to the thickness of a cigarette paper.

With the finest of emery, with the most delicate of delicate touches I again and again touched this spot. It was delight to see that scum spot growing smaller and smaller, its edge breaking into tiny spots, those spots vanishing under another touch of the emery to liberate colour underneath that glowed and laughed and danced to the light of the sun.

The strain of facing such a ticklish gem is too great for the eye; it must be rested lest sight play tricks. The colours "run" together, brighten out and fade and brighten again, appear to have greater depth than fact, the strained sight deceives itself. Facing under such conditions is liable to make a man use a turn too much of the wheel and the colour really does vanish for all time. Then, one just about cries.

But the thrill, the laughing triumph, when one gets the last trace of scum off the little gem and polishes it on the felt wheel with red jeweller's rouge. We gazed at the finished gem and it glowed back and danced in quivering flame and orange and green to the kiss of the earth's dull light.

Our little parcel faced fairly good stones. But there was only one gem, alas a tiny one, just a ring stone. But its value was almost that of half the parcel. It was worth £30 then; today it would realize more than a hundred. What the Ratters had robbed us of was worth many hundreds.

We faced all the stones except five. These were "double-bar stones". That is, two bars of colour ran through each stone, with a layer of potch separating them. Such occasional stones were the subject of much discussion on the field. If only they could be cut or sawn in half, that is sawn through the band of potch which separated the colour, then each stone could be converted into two. McIntyre, the cutter, was one of the first men to find these double bars of colour in stones; and he had imported a special machine to cut them.

We agreed he should try his hand on these five stones.

His camp was away on the edge of the Flat among the trees. I took the stones to him, and left them there, never dreaming of the direful consequences.

29. ST PATRICK'S DAY

We sold the faced parcel to the big buyer, E. F. Murphy, for £72 – if memory is exactly correct. This was a tiny parcel; Murphy was buying in thousands. But it was a thrill to sell even such a small parcel. A little crowd of us lucky ones would squat yarning in front of the buyer's bark hut, comparing our parcels, seeking to come to a closer estimate of their value. Most of the parcels were held in shabby old tobacco tins, the lovely stones jealously wrapped in black cloth. On that day, those shabby old tins held more than £25,000 worth of black opals.

Examining numerous parcels thus from various parts of the field was to realize the infinite variety of opals. Even in one claim the stones would vary – even in one patch. To see collections of opals thus won over miles of country was an eye-opener, even to the man used to opal-mining. As the colours and patterns vary vastly even in one tiny localized patch of opal, these opals taken from many patches from various claims over miles of country held unbelievable combinations of light and pattern and colour. Again and again throughout the years, we would wonder at this infinite variety of pattern and colour, and how nature had produced it. No one knows the mystery of the opal. Science only guesses.

Inside, Murphy, a quiet, black-bearded man with a kindly smile, sat at a table. The room was light yet dark, so lighted that the light was evenly distributed everywhere. From tobacco tins I spread our little parcel of opals on the table. Quietly Murphy examined them, picking up stone after stone to turn each around while gazing at it then placing it back on the table. When he'd finished the stones glowed there in little groups. He had classed them as a buyer classes opal – for some particular market he has in mind.

"How much?" he murmured.

"Ninety pounds," I answered. And looked at him.

"Just a little high. I suggest you start at £60."

"They're good stones!" I protested.

"Yes – of their class, quite good. I think we shall do business. But the price is a little high."

"Well then – £82."

He had glanced once only at the gem stone. It lay on the table in a sparkling prettiness that shamed the others. I thrilled to the beauty of the little gem that Tom and old Dad and I had dug alone. And realized that this little gem would sell the parcel.

He only examined the firsts. "I'll give you £60," he offered.

"Oh, no! They're worth more than that. Why! That gem alone is worth almost as much as that!"

"Almost!" he smiled, "not quite. It is a nice little stone but you know it is only a ring stone. Come now – £60."

"I'll take £75!"

Slowly he shook his head.

"You're just a little too high," he murmured. "I think I might place this parcel with the aid of the little gem stone, but the other stones are really not of an exceptional class. I'll make it £65."

Then I was sure of a sale. Tom and Dad and I had agreed on £65 as bed-rock price; we hoped to get £70. Eagerly I bent over the table, explaining the virtues of the stones, growing confidently enthusiastic. This was a great game this selling, as thrilling almost as digging the opals. The buyer leaned back and smiled.

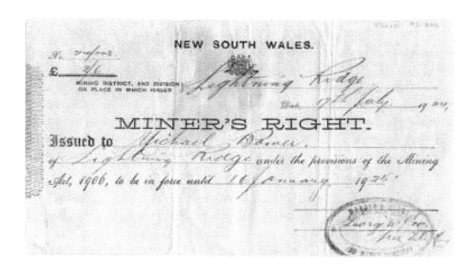

"I've a number of parcels to look through to-day," he said, "large parcels that will take up a lot of my time. I suspect your price is £65, but I'll give you £68."

"Seventy-three!" I bargained.

"Couldn't you really come down to seventy?" he smiled and reached for the pen. When his hand opened the cheque-book I stood up and grinned:

"Yes, Mr Murphy, seventy-two pounds. Make it in three equal

cheques please; one to Tom Peel, open one to our old mate we call Dad, and one to me."

He hesitated a moment, then smiled: "Oh, very well. But you're the hardest customer I've had to deal with today. I'll be on the wrong side of the ledger this afternoon if I'm not more careful."

I grinned, pridefully accepted the cheques, and walked out. I'd made my first real opal sale. Tom soon got merry with his cheque. It didn't matter; he had earned a spell.

The shaft was well chambered out now, and there was room below in the drives to throw the mullock. The claim could be worked for a week and more, Tom could pull the dirt when he recovered. The old man simply sat on the dump all day.

The scrub was falling round about now. Holes were going down all over the Hill; we could see right down the Gully and across the Flat. The hundreds of tents down there looked like shadowed mushrooms under the trees, the scattered bark huts, and Ted Dorrington's and Watty Vawse's stores appeared like massive ant-beds beside the dusty track.

Jack Prentice's butcher cart would be trotting along. Jack was a good bloke. Fairly often men arrived at the Ridge with only a very few pounds. They'd camp quietly and live on bread and dripping, toiling from daylight to dark hoping to strike it before lack of money forced them to walk from the field. When Prentice saw a new camp among the trees, the owner, if he did not order meat within a few days, would on returning to camp find a bag of meat hanging from his ridge-pole.

I remember two mates who had only £8 between them. They camped right out in the scrub, where no one could see their camp, and lived on bread and fat for six weeks. When Jack Prentice found out, he was very indignant – wanted to fight them both.

Very, very occasionally, one of those hard-ups would be lucky. There was Bob Austin for instance. With his belly crying from hunger he went down an old abandoned shaft. An hour later he was worth £1200. Such lucky ones were few and far between.

One day we received a surprise. Climbed up the shaft to hear the thud of pick blows. Bill Bullock and George Pedley were starting a hole within a few feet of the shaft. I stared.

"Didn't think you would leave this bit of ground unpegged, Jack," smiled Bullock from their windlass.

Tom and I (and for the first time) without a word walked to the pegs and closely examined them. There was no doubt about it. Old Dad had pegged the claim but he had only just pegged in the shaft. Actually it was only twelve feet within our boundary.

"We did a lot of staring ourselves," said Bullock, "before we were sure. But that strip was vacant ground all right.

"There are the adjoining claim's pegs, and there's yours, with this narrow strip of ground vacant between. When we were sure there was no mistake about it, we put in our pegs."

"Good luck to you," I murmured.

"I'll go hopping to hell," said Tom.

Old Dad hung his head. He had blotted his copy-book disastrously. First, his tongue had put the Ratters wise and they had robbed us of hundreds, possibly thousands of pounds worth of opal. And then he had fallen down on the only simple job we thought he had done. Instead of the proved shaft being in the centre of the ground, the shaft for which we had risked the ridicule of the field, he had pegged out on the very boundary. And the opal patch might well run out through that boundary. Instead of us having an opal claim, it might now well prove a duffer. Glumly we worked, watching Bullock's and Pedley's shaft sinking deeper day by day.

They bottomed at about eighty feet, on good potch and colour. In a few days they were on opal, an extension of our own patch, we later proved, that had run out under our boundary. We were toiling solidly, and not getting one solitary stone.

Bullock and Pedley drove towards us, we drove towards them, and both parties gouging, broke through on the boundary. Just a small, round hole deep below the earth connecting the drives and the two shafts. We could poke our heads to the shoulders through that hole and talk. Thus each party knew their boundary, measuring below from each shaft, along the roof. We marked a dividing-line and worked on – Bullock and Pedley on opal, Tom and I on none.

Tom broke out on the hops again. No one could blame him. We had slaved to sink that apparently worthless shaft, and through no fault of ours we had been robbed by the Ratters, and now we had lost the opal ground. While Tom was away I worked doggedly on and mullocked up both drives and shaft. Then was practically powerless to do a thing; there was no one to pull the mullock and thus clear the shaft and drives. Tom was well set for a long drawn-out bender.

The inactivity, with so much activity around, was maddening. Old Dad tried his hardest to do something. He manned the windlass, and pulled up a few buckets of mullock, morning and afternoon. I'd fill a bucket, then wait while it went slowly, so slowly, up the shaft. Old Dad's gasps would come down the shaft now and again. Then half-way up the shaft the crawling bucket would stop while he had a spell; then it would creep up another ten feet, only to come to a stop again. I'd creep into the drive lest the old man's grip on the windlass handle slip and the bucket come crashing down the shaft. They were only the usual iron oil-drum buckets – not the big bullock-hide buckets that would hold three times as much. With bullock-hide buckets and a rope with two hooks I could keep any man busy for hours, one bucket down the shaft being filled, one full bucket being hauled up while the empty bucket passed it coming down the shaft. It was only to spare the old man's feelings that Tom and I used the small iron buckets.

The diggings at Lightning Ridge 1941.

But Dad just couldn't work, and that was all about it. Day after day I had to sit on the shaft with him, waiting for Tom and his friends to finish their spree.

I'll never forget St Patrick's Day. The shafts were deserted, for every one holidayed, except those lucky ones on opal. Bullock and Pedley were working below in their shaft; old Dad and I sat on our dump gazing down the Hill across a forest of moveless windlasses.

Just after midday I exclaimed: "It's no good, Dad, I'll go crazy sitting here day after day doing nothing. We'll have to get this shaft cleaned out somehow."

"You won't let me go down," he mumbled.

"You know you can't do any good, Dad."

"Let me go down and fill the buckets then. I can fill quicker than I can haul."

"Oh very well. Give it a go; it will be attempting something anyway."

So he slipped off his old shirt, sat on the shaft mouth with his legs dangling, and placed one foot in the loop of the rope. His skinny old arm reached out for a grasp of the rope.

"Do you think you can manage it, Dad? After all, the afternoon will soon be half gone. I think I'll go for a walk instead. Never mind, Dad, don't go down. Tom may steady up by morning."

"You know he won't be fit for work for days," he quavered. "An' I'm goin' down! You won't never let me do nothin'!"

He put his feeble weight on the rope. "Right!" he said.

Slowly I lowered him, thinking what a light weight he was, only skin and bone. Presently he came to rest on the shaft bottom.

Slowly he lit the candle, groped about to fix the bucket, slowly began to fill it. Away down there, illuminated by the candle-light the white of his bent old back was vaguely visible.

Very, very slowly he filled a few buckets, I would lean on the windlass and wait, sometimes hear his faint gasps coming up the shaft. After an hour of it I shouted to him to knock off, to come up. But his quavering old voice called back:

"I'm all right. Send down another bucket."

Time dragged on; he was toiling slower and slower. He'd blown the candle out when his eyes got used to the gloom. I grew uneasy when he faintly gasped occasionally. It would be pretty awful should he die of heart disease.

"Come on up, Dad," I shouted. "I'm knocking off."

"Just a few – more – buckets," he called back.

30. DEATH

Across from the camps a figure appeared, walking unsteadily. It was no pleasure to watch him approaching. He was one of Tom's drinking acquaintances; no particular friend, he merely joined in amongst the revellers whenever they "broke out". A ship's steward, he had been attracted to the field by rumour or fate or whatever is responsible for the happenings in the lives of humans. A decent fellow when sober. At the camp-fire he had told us all about his wife and two babies waiting for him in London. He was very proud of them; they were proud of him; they were sure he was going to return a "big" man. He toiled in hopes of striking it rich and surprising them at home with "big money". But he'd been unlucky so far. And now he came staggering among the bushes, unsteadily coming towards the claim.

Dad had been an awful long time filling these "few more buckets" of his. I leaned over the windlass and called: "How are you going, Dad?"

He didn't answer. I grinned in sickly fashion, looking out over the windlass, watching the steward approaching. Dad was having an exceptionally long spell; maybe he was sitting in the drive cooling off. Certainly he heard that shout but as the steward reached the claim and began climbing up the dump I shouted again down the shaft: "How are you doing, Dad? Make that the last bucket, Dad, and we'll knock off."

No answer. He didn't want me to know he was spelling. As if I couldn't tell! Oh well, let him spell as long as he liked. I turned to yarn as the steward sat down.

"Why don't you come down to the camp and join in with the boys, Jack? Come on! What's the good of toiling up here! You've no need to."

"I'm working," I answered shortly.

"But it's a holiday, St Patrick's Day."

"It always seems to be a holiday, where you and Tom and your particular friends are concerned. One long holiday."

"No need to rub it in. The boys need a spell sometimes. Who's down the shaft?"

"Old Dad."

"That old swine, I can't stand him; he insulted me the other day."

"Your fault. Dad doesn't say a word to any one unless they speak to him. And he always answers civilly – to a civil question."

"I detest the old swine!" he exclaimed, and snatching a lump of opal dirt, hurled it down the shaft.

"My God! You drink-crazed fool!" I shouted. "You might have killed the old man!" I knew he hadn't. Opal dirt is generally soft, and I'd seen this clod strike the side of the shaft and splatter into fragments as it went down.

"You're a b— fool!" I snarled. "Get to hell out of this back to camp with your drunken mates." I was listening, expecting Dad's quavering voice for he must have been sprinkled with the little shower of pellets.

At last I called out:

"Dad!" He didn't answer. While listening, a slow, cold fear gripped me. I leaned over the shaft and shouted:

"Dad!" Silence. I stared around into the steward's questioning eyes.

"What's the matter?" he whispered.

"You awful fool," I whispered urgently. "He's not answering."

"He *must* be. Shout."

"A dead man would hear me! And my God he doesn't answer." I shouted again.

"Quick! Let me down this shaft!" He sprang for the windlass handle; I slipped a foot in the loop and swung out over the shaft.

"For God's sake steady! Don't let me crash down! Right!" He lowered me with dizzy speed. Almost instantly I saw away below the white of the old man's back lying on the shaft bottom. There was sweat on his back, caked in powder of opal dust. I spread out both legs and landed with a thumping jerk, one foot on either side of the old man's body; he took up nearly every inch of the narrow shaft. He sprawled there face down, his old head wedged against the bucket. I touched him and shuddered at the clammy feel. "Dad! Dad! Dad!" I whispered urgently. Away to the left down the drive was a faint glow in the gloom.

"Bill! George!" I shouted. A candle-flame, then Bill Bullock's face appeared framed in the little hole which connected our drives.

"Come quickly, Bill! Something's happened to old Dad. He's collapsed – or something."

A candle was thrust through, then both of Bullock's arms, then his head as he tried to squirm through the narrow hole.

"Push!" I heard him say and pictured broad shouldered George Pedley pushing with his powerful strength. Bill suddenly squeezed through, clothes and skin torn. He crawled down the drive, then bent over Dad.

"Roll him over – gently," he said. Together we managed it, a difficult job in the narrow space. Bullock put his hand over the old man's heart, then held his ear there. He wet his finger-tip and held it to the old man's nostrils.

"Dead!" he said.

I heard the blows of Pedley's pick as he swung to widen the hole so that he could squeeze through.

"Run to the camp," I shouted up the shaft. "Bring whisky or brandy or something." The steward's head disappeared like that of a startled rabbit. George Pedley broke through the drive and came crawling along to the shaft.

"He seems dead," said Bullock. "But we'll massage him, warm him if possible until the brandy comes. I'm afraid it's hopeless though."

"It was a nice way to die," said Pedley, "clean and swift."

Kneeling there, with their heads and shoulders out of the drive they vigorously rubbed the old man while I stood helplessly against the side of the shaft. There was only room to stand still and wait.

"Below there!"

"Hullo."

"Here's the brandy!"

"Right. Tie it securely to the rope, and lower it."

"Right! Here she comes!"

A sound like roaring wind, a thud as the full bottle of brandy crashed beside Dad, falling between the bent heads of Bullock and Pedley.

Across all that shaft, there was only six square inches of dirt not covered by the bodies of old Dad, of Bullock and Pedley, or of myself. And right on to this tiny space that bottle dropped from a hundred feet. Dad was fated to die that day, but none of us three. We stared at one another. My heart started thumping again.

"You crazy fool!" roared Bullock up the shaft, "you might have killed us. You b— drink-crazed fool!"

"It slip-slipped out of my hand before I could ti-tie it!" stuttered the steward's voice. "The string wouldn't tie – it's com-coming now – look out!"

As the coil of string came floating down the shaft Pedley roared: "Get to hell out of that, you crazy fool. Call the boys and clear out before I come up and murder you!"

"That's what you could call a miracle," said Pedley soberly, "that bottle crashing just there instead of splattering our brains out. It didn't even hit the dead man."

"Just as well," remarked Bullock, "or they might have thought a stone had fallen on him."

"You'd better go up on top, Jack," said Bullock. "You're a bit white about the gills. I hear the boys tramping up on top now. Go up above, lad, we'll fix everything down here."

I started to climb. Then he said: "No Jack. Crawl through the drive and climb up our shaft. There'll be no chance of a stone falling down then. It would be sheer stiff luck if something fell and hit the old man's body. Even a pebble might leave a bruise like a blow from a hammer."

It was good, cool advice. I was feeling white, but it was at thought of the steward. I knew the clod of dirt hadn't struck the old man; I'd seen it fly into a score of little pieces. It was only soft opal dirt too. Besides, I remembered the old man hadn't answered when I called down the shaft before the steward arrived. Old Dad had died when I thought he was having a spell. It was a spell right enough.

There was a crowd up on top now; they were coming too from their camps away down the Flat. Tom came up the dump. He was very shaky, his eyes were troubled: "What's happened, Jack?" he asked quietly.

"Old Dad. He died down the shaft." He turned away and I stared into the frantic eyes of the steward.

"You won't tell!" he gasped.

"What do you mean?"

"You know, Jack! I didn't mean to do it! It didn't hit him really! It couldn't! My wife, Jack – my two kiddies ..."

"You'll have the men wondering what you are talking about," I murmured.

He slunk aside. They were preparing to haul up the body. Many of the gougers came and quietly sympathized with me. The steward stood by, his ears wide, his eyes imploring every time I opened my mouth. The fellow's almost hysterical fear was beginning to haunt me, to make me think of that stone. Though I was sure the old man had been dead ten minutes at least before the clod shattered in fragments against the shaft.

There was a big crowd there by the time they hauled up old Dad and laid him under a tree. If I opened my mouth the steward would swing from that tree – at the end of the windlass rope. Over the heads of the men his eyes stared at mine. Terror glared from those imploring eyes.

"Well, I'm damned!" exclaimed a gouger. We turned and there, riding up along the red track across the Flat was a mounted trooper. What was he doing here? This day of all days? He turned his horse towards the big crowd on the Hill.

"Oh well," said Bullock, "he'll take charge now."

He did, a tall young trooper, stern, sharp, efficient, quiet. He took all particulars from me, then from Bullock and Pedley, noting everything in his note-book. His questions were very direct, very much to the point. Apparently satisfied, he turned to me very quietly: "Were you alone on the dump?"

In a flash I remembered the steward's wife and babies. Old Dad was very, very old. The babies were very, very young.

"I was alone," I answered. Without seeing him I knew the steward had turned around and was stumbling through the crowd.

After the trooper had had the body taken away, I walked slowly back to camp with Tom. Gradually the crowd dispersed, leaving the Hill to the afternoon shadows.

Tom sat with bowed head, the shock was sobering him. Many things were flashing through my mind. Especially how jolly good it was now that I'd been honest, right to the last shilling, with the opal. I'd had charge of it; had faced and sold it. And given Tom and Dad their shares to the last shilling. The young trooper had been so careful in noting all details of the opal we had won, and sold. And then:

"My heavens! ... Those five bar stones I left with McIntyre!"

Abandoned school, Lightning Ridge.

31. DEADMAN'S CLAIM

Tom stared from watery eyes.

"What about them?" he croaked.

"Those five stones with the double bars that I couldn't cut. When the policeman asked me particulars of the opal we'd won I told him everything – except those five stones. I completely forgot them. Don't you remember he pointedly asked?

"'And are you sure you have told me everything; that you've accounted for everything; that you've got no more opals at all that you've forgotten to account for?'

"And I answered: 'We haven't got a quid's worth of potch and colour even, in the camp or anywhere else.'"

"Blast him anyway," exclaimed Tom. "Don't worry about that. It's nothing."

"Blast him too," I answered. "If I went looking for him now, and told him about those stones, he'd think we had others planted away. Anyway, those stones have nothing to do with him, or the government either. Old Dad had no relations; the government would only take his share of those five stones. I'm glad I didn't tell him."

"So'm I. Don't say a word about it. It's nobody's business but our own."

"It's a rotten business altogether, quite apart from old Dad. I wish to goodness you'd sober up and help me. Here you are a solicitor; and here am I, ignorant of law, but right into it and forced to take everything that's coming."

"Don't reproach me now, Jack," implored Tom, and jumping up strode away.

I walked away too, out towards the bush. But soon, something turned me back towards the Hill. I walked up among the leopard woods and sandalwood and bushes towards the claim, thinking and thinking and thinking. The sun was sinking fast, the hill-top was red. Suddenly I stopped, scared. The trooper's horse was tied to a tree near the dump.

Why had he doubled back? We all thought he'd gone to the Old Town with the body. Creeping amongst the bushes I peered towards the dump. There was the trooper, right on top of the dump, outlined by the rays of the sunken sun. He upended an iron bucket and knelt, examining the bottom ... Searching for hairs or bloodstains on the sharp rim. He thought I'd dropped the bucket, or lowered it too fast down the shaft on to old Dad's head!

It was a horrible feeling, realizing what that trooper was thinking, what he was looking for. He examined both buckets then, carefully testing the windlass, lowered a bucket right to the bottom, carefully noting just how it glided down the shaft, then hauled it up swiftly, then lowered it very swiftly. He was testing to see whether, by swiftly lowering the bucket it was possible to knock out a stone from the side of the shaft! I knew that the windlass was set too true for that. Then he examined the rope, particularly the iron hook on the end of it. Then handled the windlass, and examined the logging of the shaft, testing if the spaces between the logs were so protected by bark that a jar from the bucket would not cause a stone to fall through. Then he stared down the shaft a long, long while.

He was memorizing that shaft, trying to reconstruct exactly what had happened: Had the old man died naturally? Had I killed him accidentally? Had I killed him purposely?

My heavens! If he had known what the steward did!

He probably was thinking: This was a new claim, a rich claim which might contain many thousands of pounds worth of opals. The old man was very old, and friendless. Might not his two mates have conspired to put him out of the way?

What if there was a mark on old Dad's body?

The trooper, a black silhouette now, stood up, and stared slowly around. A whirl of thoughts tortured me. That trooper might be visualizing that *I* threw a stone, a real stone and not a clod of opal dirt, down the shaft. And could he find the faintest evidence it would be I who would have to prove I didn't do it! The steward was out of it; I'd said I was alone on the dump. Tom was out of it; he'd been drinking with friends for days past – and wasn't near the claim. Bullock and Pedley were working beside us, but underground, and they couldn't enter our claim until they'd broken the man-hole large enough. And they hadn't done that until I called out to them. No, there was only "Jack" Idriess.

Darkness came, and the trooper was still there. Stealthily I crept back towards camp. A figure stepped from the bushes.

"Don't speak to me!" I snarled.

The steward slunk miserably behind while I walked across to old Matt's fire. A crowd of the boys were there, quietly yarning about the events of the day.

The next two days and nights brought desperate worry. Would a mark show on Dad's body? In imagination I saw that policeman examining the body as carefully as he had examined the shaft. Dad might have hurt himself when he collapsed, pitched forward on his head against the end of the shaft, or the pick. Bullock had mentioned that even a little bruise shows out big after a man is dead. I'd never have thought of such things had I not seen the trooper examining that bucket. And I daren't breathe a word to a soul – not even to Tom, the man of all men whose position in his earlier life could have helped me. He was almost maudlin now, blaming himself for Dad's death.

The steward never let me out of his sight. His eyes followed no matter where I went. And I was bushman enough to discover that at nights he came crawling to our tent and lay just outside, listening. He was a fool, his mind so filled with fears he didn't realize that now I daren't say anything. Even if a bruise was found on the old man's body and suspicion arose that he had been killed with a stone, I daren't say a word about the steward. For they would say:

"But you told the policeman that you were alone on the dump!"

Every moment seemed a month to the steward and me, waiting until the old man was buried. Tom drowned his sorrows. The burial day dawned at long last. Tom and I trudged down the track to the Old Town. We'd walked several miles when I glanced around. The steward was following, away behind. At the newly built hotel we breasted the bar and had a drink. Old Doc and a few of the boys were there. Old Dad was lying somewhere inside. The trooper presently appeared and called the doctor and myself. He motioned back Tom and the impulsive steward: "You're not wanted!" he said.

Doc and I followed him into a dim room. Dad was lying in the box. The Doc pulled back the sheet, examined the body in a businesslike way, then with thumb and forefinger pushed the head around. My heart ceased beating and the policeman leaned forward too, but the doctor turned the head back and flipped the sheet over the body.

"Thank God!" I thought.

I'm not positive about Burke but felt that both he and I stared in our different ways expecting to see a bruise on the temple. Imagination is extraordinarily powerful when fear works upon it.

"No doubt about it, Burke – heart disease," said the old Doc testily. "Come on." The trooper demurred. But the doctor snapped: "Box him up!" and stumped out.

I followed him – couldn't get out of that place quickly enough. The dead man lying there stiff and cold might have sharpened an uneasy mind. I felt intuitively that the trooper had voiced some suspicions to the doctor. When we entered the bar again the steward sidled up and whispered, "Have they screwed the box down yet?"

"Go to hell!" I hissed.

The trooper did not appear. I had a chilly respect for that young constable; he was so keen, so coolly efficient; a man would have been into terrible trouble indeed if anything really had happened to old Dad, even had it been but an accident.

They put him in a box, shoved the box on a dray and away we lumbered. The trooper riding ahead, then came the dray. creaking to the slow old horse, then Tom and I walking slowly behind. No need to glance behind. The steward was there; he must have been nearly singing now that the old man was at last in the box. Only a little time now, and he would be out of sight for ever.

The hole was a red scar in a restful spot, big trees all around. There was sweet green grass, and clinging to a drooping mistletoe a bird sang as we lifted the box from the dray. We were the only mourners.

We lowered him and pulled up the rope. Then the trooper read the burial service, the first such service I'd ever heard. Then he threw a handful of red earth in. Then the drayman filled in the hole, and we went away.

The steward chattered beside us back to the hotel. His footsteps were springy; he began to tell a laughable yarn. But Tom suddenly turned on him. When we reached the hotel Tom jumped up on to the veranda and hurried into the bar. The steward turned to me: "Thank God it's over," he sighed. "I've been through hell. And now I'm going to get so drunk my own mother wouldn't know me."

"And talk!" I sneered.

He stared, then his face went white: "I'm going back to camp," he whispered, and turning abruptly walked swiftly back along the track.

I sauntered into the bar after Tom. He and the old Doc were starting on a great binge.

I left them there that night and slowly walked the quiet miles back to camp and sought out the steward. "You're going back to England, I believe?" was all I said.

"Straight away. I've had enough of this field to last me a lifetime." And he went. Some few months later he posted me a photo of his wife and babies from England. I was glad then I'd said nothing.

Tom and I could not work the claim for some little time after Dad was gone. The trooper impressed upon us that the Law was entitled to sell the third share in it, on behalf of the dead man's relatives, if any. Indignation spread over the field. In the eyes of all, that third share no longer existed, the claim was Tom's and mine. But the Law had come to the field now. Even so, the gougers all declared that Law or no Law no third party was to "butt in". Deadman's Claim now belonged without question to Tom and me. So that on the day of sale the mob rolled up, a black mass around the dump. Every man knew he was there not to bid.

The trooper sensed it. He mounted the dump, a tall, straight young figure, very determined-looking. He was purposely slow in getting his papers ready then he faced us all and made a short, clear-cut speech, explaining the mode of procedure. Then he described the claim and made a pretty picture of it, pointing out the wealth of opals that must be awaiting the lucky purchaser in the unworked ground of this valuable Prospecting Claim, now known far and wide as Deadman's Claim. Then he called for a bid.

"How much now? How much am I offered?"

Not a murmur. He called again. No movement, just a sea of upturned faces. He gazed around, lowered his hands, squared his chest.

"Now men, you must start me with a bid, the condition of this sale is that it *must* be completed to-day. This is your only and final chance, a chance for a probable fortune. You know even better than I, that this unworked ground must be very valuable." And he described the claim and the opportunity again, very convincingly.

"Come now!" he shouted briskly. "What am I offered for the third share in this claim?"

A short, fidgety man with big eyes and open face had gradually pushed himself right to the foot of the dump, gazing up at the policeman-auctioneer. I saw what was coming. This was the man, you'll meet him in every crowd, in every meeting, who does not know what he is in there for. This was the one man among all those hundreds who did not know he was there *not* to bid. When the trooper called for a bid again, this man shouted a bid. Instantly a nasty murmur started, rippled down through the crowd. Angry shouts rang out above the murmuring. The bidder glanced around in bewilderment, then exclaimed.

"Why! I thought this was an auction sale! I withdraw my bid!" he shouted up the dump and backed away in amongst the crowd.

Frowning at the quick, sarcastic laughter, Burke stepped smartly to the edge of the dump, held up his hand, and called: "Men! This is an official sale and every man here is under the protection of the Law. I'll allow no bidder to be threatened! Every intending bidder here is under police protection, now and in the future too!" He spoke up well.

Tom, standing with folded arms, laughed sarcastically. I was exultant, while just a bit sorry for the trooper. He was putting up a real game case, but his chance was hopeless. No doubt to have got a big price for that third share would have been a feather in his cap. But that share was Tom's and mine. I thought of all the hopeless labour we had spent in developing this claim, working here when all men thought us mad. But for us, probably the place would still be covered in scrub, instead of a scene of intense activity, already turning out thousands and thousands of pounds worth of opal.

Again the trooper called for a bid. There was a tense silence. Again he called.

"One pound!" I shouted, and smiled up at the dump. The trooper walked around up there on top, again he called out the virtues of the claim, and how every one was free to bid. Again he called for a bid.

"One pound!" I called. He started off again, but it was no use. At last he had to knock it down to Tom and me. We bought in our late partner's share of Deadman's Claim for twenty shillings.

Wembley Exhibition, 1920, showing the world's first Black Opal.

32. THE TROOPER

But a few days later, it was the trooper's turn. He came riding up to the camp at sundown – I liked the man but had come to dread the sight of his well-groomed horse, the trim figure of him riding up the red, dusty track. He called me away from Tom.

"Mr Idriess," he said, "you told me that you had accounted for every pound's worth of opal, and for every stone that ever came out of your claim – Deadman's Claim!"

"Yes. That's right!"

"Then – why did you not tell me of the five opals that you left with Mr McIntyre?"

I just stared at him. He stared back. "Well?" he asked sharply.

"It's this way," I answered slowly. "Those five stones were bar stones, I could not face them. I left them with McIntyre to be cut. But all the other stones, every shilling's worth of them that we got, I have told you about."

"Why did you not tell me about those stones?"

"I clean forgot."

"How do you expect me to believe you when I have found you out in this deliberate deception?"

"Well, it's this way," I answered slowly. "I don't know. And – I don't care!"

"That is an answer I do not like. Now why, really, did you not tell me of those five stones?"

"I forgot in the first place. Then, when I did remember them, I came to the conclusion that the government had no right to those stones. And neither it has. We dug those stones! We mined them! The government didn't!"

"That is not the position at all. One third share of those stones belonged to your late mate's next of kin, if any. If not, then to the Crown!"

"Well, I'm right against you then. The old man had no relatives that we know of. Whether or no, those stones would still be deep within the earth if it had not been for the sweat of Tom and me."

"That is not the point at all. Now – are there any more opals that you have not told me about?"

"No."

"Are you certain?"

"Yes."

"I do not believe you."

"Please yourself."

"Right. I will."

I walked back to the fire. He mounted and rode away, his sharp eyes all around the camp, no doubt thinking about digging up the camp and searching for buried opals.

"Those stones!" I shrugged to Tom, "those five double-bar stones that I left with the cutter."

"My heavens! Has he ferreted *that* out?"

"Yes."

"Blast him!"

We started work on the claim again, toiling hard. Driving side by side with Bullock and his mate along the boundary, it was wonderful, but utterly tantalizing, the way the stones clung to their side of the boundary. Day after day, night after night, there would be that line of opal along the face, ceasing at our boundary as if it had been cut off with a knife. One evening Tom put down his pick and sat back with a sigh, perspiration gleaming on his body. It was warm down there; the candle burned very low.

"Dad got out of the claim in time," he murmured and rolled a cigarette.

"Yes. A new claim – a Prospecting Claim – and it almost seems to have duffered out! Can't understand it."

"Just as well we're not superstitious," he murmured.

"Why?"

"We'd say the old man had put a hoodoo on the claim."

"Rats."

"Sometimes it almost feels as if he's watching over our shoulders and driving the opal away."

"You haven't been drinking?" I said slowly.

"You know I haven't tasted a drop for a fortnight." We smoked in silence. It was deathly quiet down there. Tom's heart was softly pumping.

At last the patch spread a bit, a little of it came over the boundary into our ground, and we had the thrill of being on opal again. We'd "chambered" the boundary between the two claims now and at night worked side by side, the four of us in line sitting on our knees carefully "chip! chip! chipping" at the face. Now and again Bill Bullock would chip very delicately, then lay down the pick and hold the candle to the face. When he sat back we'd say:

"A stone, Bill?"

"Look see!" he'd reply. We'd crawl over and stare at a dull red bar glowing from a nobby in the face.

"Looks a big one!"

"Flame bar! If it goes all through, it's a gem."

"There's another one beside it! I can just see the potch in behind there."

Very carefully Bullock would gouge under the stone, then lever it out with the long, thin point of the spider. We'd kneel there breathlessly as the nobby broke away from the face. With strong fingers gentle as a girl's, Bullock would turn the nobby around under the candle-flame.

"Look! the bar goes right through! Red!" And we'd see the dull glow deep within the smudgy potch. Bullock would pick up the snips. Click! "Ah!" And flame would flash where the jaws of the snips had bitten a chip from the potch. Another snip on the opposite edge of the nobby and "A gem!" Orange and flame and green danced under the candle-light. A mysterious glory, a miraculous thing those pulsing colours of light imprisoned in jet-black potch buried deep in the earth. Those colours seemed living things when they leaped to the light after the liberating jaws of the snips.

Tom and I did not get much; only the tail end of the patch where it overlapped into our boundary. But it was enough for a holiday to Sydney.

I stayed with old gran. On various "spells" in Sydney throughout the years she'd receive me with tears, but was quite dry-eyed at departure. Evidently my peculiar charm was of a type that fades on closer acquaintance.

The sisters were at boarding-school, and growing up, a habit of children with the passing of the years. Dad was still away in Broken Hill, soon to be transferred to Grafton.

It was a pleasure to bump and rattle and puff into town on the old steam tram and view this city from the eyes of an independent man. Acknowledging no master, watching from a height these swarms of citizens rushing to obey the clock. The tiniest turn of fortune's wheel and I, too, would to-day have been one of them.

A further turn of that wheel would have landed me a bloated plutocrat. For adjoining gran's old house at Waverley, where those rows of massive cottages are to-day, was then a paddock, the stabling grounds of a line of horse buses and cabs. And Bondi and Coogee were tiny villages isolated in deserts of sand. Such places as Maroubra, Sans Souci and Cronulla, etc. were merely names, associated with the wreck of the *Hereward*. I often remarked to gran that that old paddock would be valuable one day; that I'd like to buy it, and some of those acres of sand on the sea front at Bondi and Coogee and Waverley, too. She offered scant encouragement and I lacked the initiative for business action. Unless lad or man or woman put their plans into action then the richest schemes are vain. Strangely enough, the old gran was a shrewd business woman and had made thousands by dealing in property.

One day I strolled down to the Sussex Street wharf, the old *Newcastle* was in. I had a tobacco tin. Within it was a little strip of black velvet on which were glued bright little opals, chips of stones and specks of potch and colour shaped into ring and tie-pin stones. All hands on the Ridge thus fashioned their little chips as ring stones, mostly to give away to friends.

I yarned with the boys on the *Newcastle*, and gave away a stone or two. When leaving the wharf, one of two strangers accidentally bumped me. Confidence men.

[I'd often wondered at the transparent motives of most of these fellows, and why they always thought a "bushy" in obvious clothes was so simple. Wondered too, just how they could possibly succeed in their business. These two well-dressed thugs, prowling on the wharf, had noticed the seamen bending over the little tin of opals. And the shrewd fools thought those little stones valuable.]

After apologising: "Gosh, mate, it's hot," he said, and wiped his brow. We strolled out of the wharf gates.

"We're strangers," remarked his mate, "just down from Dubbo. Where can a man get a drink in this outlandish place?"

"There's a pub across the street," I nodded.

"Care to join us, mate?" invited the other, "a little one won't do us any harm."

"Right-oh," I agreed, wondering just how they would try to take me down. We stepped into the frowsy hotel.

"What's yours?" one smiled invitingly. He was just the ordinary, low-browed thug. The other was taller and strongly built with a face that obviously hurt him to smile. The barmaid poured me out my beer – from a bottle.

"Beer for mine," they ordered. But she pumped their beer for them. "Here's luck!" they greeted, and raised their glasses.

"I'll have one out of the pump too," I said brightly. "I don't like it out of a bottle."

"You ordered it," snapped the barmaid, "what's wrong with it?"
"Nothing. But I want it out of the pump – it's a fancy of mine."

She hesitated, then pumped it out. Other lounging customers were drinking similarly; they merely glanced at us no doubt thinking the confidence men had caught another "mug".

The smaller thug nodded to the barmaid, she "filled 'em up again".

Innocently I watched her.

"How's things way out Bourke?" drawled the tall thug.

"Dunno."

"I've seen you somewhere outback," remarked the other – "Narrabri, I think."

"No! It must have been Moree. Weren't you working with Jimmy Smith and the gang?"

"No."

"I've seen you somewhere, could have sworn it was Narrabri – must have been Boggabri. Can't just place your name."

"Might have seen me in Narrabri. When was it?"

"After the shearing."

"What season?"

"Oh, it might have been last season ... or the season before."

"Wasn't me."

"I'm sure I've seen you there, sometime. Weren't you mates with Mick Ryan and the boys?"

"Never heard of him."

I shouted in turn. They tried hard to mention a town or an allegedly mutually known man that we could get matey about. But I asked counter questions, and the conversation lapsed.

"I think we'll have one or two in the parlour," yawned the shorter man, "more comfortable in there than leaning over this dirty bar. Fill 'em up agen, miss, please – a beer from the pump for our mate."

One led the way, I followed, the other came behind. I thought about diving out the door, but it would be more interesting to see it out. We walked down a low, dark passage, then into a gloomy room. Instantly I was wide awake. Cornered now, it meant either a cracked skull, or escape. There was only one door to the room, a small round table and three chairs stood at the back of the room below a low window the glass of which had been smeared over. I made for the window, turned around and sat with my back to it. The others sat down with a swift glance, for now they had to sit facing me. I could see over their heads to the door. Out of that, but coming down the gloomy passage was a dull rumble from the street. A man could be murdered here. If he shouted – nobody would hear him.

The frowsy barmaid came in carrying the drinks on a tray. "The beer from the pump" would be well and truly drugged.

That was how they did it. I'd been a fool. But they were placed by the table in such a position they could not knock me unexpectedly. There appeared to be a faint chance, to act the innocent then when the critical moment came to bash a chair on the head of one, to rush the other and keep rushing straight for the door, down the passage and out into the street.

"Well, here's luck."

They picked up their glasses, drank, put them down.

"You're not drinking?"

"No, not for a while. I don't like them too quick; gives me a headache."

"That little drink won't hurt you. Drink it up, lad, it's my shout," and he rose for the glasses.

"Right-oh. But not now. Get your drinks and I'll have mine later on."

Thoughtfully he picked up the two empty glasses, then turned to the door. Under the table I gripped the chair. I'd give him time to reach the bar, then crash the chair on the other man's head and run for it. But he only went to the door and whistled. Presently the barmaid came and took the tray away. He turned back and sat down.

He slapped a hand to his knee.

"I know now where I've seen you," he laughed. "Of course! At Lightning Ridge."

"Never been there in my life," I said.

They couldn't contradict, and looked rather blank. They started talking about opals; asked me if I had ever seen any stones; they'd just love to see some. They still tried hard to get into a confidential conversation with this bumpkin-headed fool. The barmaid brought the drinks, and one for herself.

"These are on me, and you've got to drink with me," she smiled brightly. "I like your boy mate, I love bushmen."

Coyly they raised their glasses, bashfully I raised mine. They drank – stopped.

"You're not drinking?"

"I'm saving it up."

"Oh, I say now, you must drink to the lady – she's shouting in your honour."

"I'm too shy," I giggled. I acted the shy bush lad in a way surely no lad would act. But it went down; I dodged the drinking. Sombrely they drank.

"Fill 'em up again, girlie."

Unsmilingly, the barmaid flounced away. I was listening hard all the time, now and again faintly hearing footsteps shuffling past the window, it must originally have opened out on to the back of the hotel.

The barmaid returned with the drinks. She put them down and went away.

"Drink!" they said and raised their glasses ... I dodged as a glass was flung at my face, at the same time upending the table against their rising bodies, and snatching up the chair crashed it through the window and followed the chair. Picking myself up on a greasy path I ran past dirty kitchens out into a filthy back yard, leaped up on top of garbage tins and sprang for a fence. Scrambled over and dropped into the street and ran, and kept running, ignoring citizens who turned to stare.

I returned to Lightning Ridge. That was the life for me. And got a shock on greeting my new mate – Old Black Joe. Tom had taken him into the claim in some expansive moment when the whisky made all men friends but blotted out the old and trusted ones. Tom had no need of a third mate. He hung his head. It was just one of those things he did. When he did it, why he didn't invite one of the young Australians who were our real friends ... well, he just shook his head.

I had nothing against Old Black Joe. But down there in the drive with him working beside me, his thick old lip hanging over his chin, his greyish-black, sullen old face with little suspicious, staring wrinkled eyes – it was exasperating. Almost any of our friends would have "given his right arm" to get into such a claim. And though the lads said nothing, I sensed the implied reproach. Not only our friends but the gougers in a body had stood firmly by us when we were in trouble over the claim; at the sale any one of them could have lawfully bid for old Dad's share. Instead, they had stood stolidly and seen the entire claim given back to us.

And there was something else. There had been some quarrel, some misunderstanding, between Tom and Joe. What it was I never exactly learned. But it smouldered actively between them so that the camp had lost its carefree comradeship of days gone by. For some ill-defined reason we worked sullenly, suspiciously.

Later, came a thunderbolt from the blue. The trooper's horse and the trooper. As he came riding up to camp my heart crept down to my boots.

It was an exhumation order, or some such technical procedure. The trooper wanted the old man's body exhumed, to be re-examined. A travelling magistrate was due to arrive at the Ridge in a couple of days. I was notified to attend the Inquiry.

As I stared after the trooper, as he rode away, Tom asked: "What does he want now?"

"Wants the old man's body dug up."

"Good heavens! What for?"

"Blessed if I know." We stared at one another.

"This is the queerest thing," mused Tom: "an exhumation order after all these months! The old man died of heart disease; there was no question about it. Why then, should they want to dig him up? ... By jove! That's strange!"

"What is?"

"I've received no notice to attend."

"You haven't all along," I grumbled. "It's I who have had to take it in the neck all the time." Tom was silent. I was picturing that skeleton ... Would it have a cracked skull? That is what the trooper must have thought.

Carving by Ron Hanlin, Chambers of the Black Hand, Lightning Ridge.

33. GHOSTLY COMPANY

The Investigation was held at the hotel, a large room was turned into a court room. Just the magistrate, the trooper, and I. Burke, spick-and-span, was coming and going soft-footed from the magistrate's room with important-looking documents. I had to wait outside until "called". It was a gloomy waiting, digging up all this old business again; wondering and wondering if they really would find anything.

At last I was called. The magistrate, an elderly man, sat at a desk neatly stacked with official papers. He was frowning, writing at a great pace, evidently with a lot of business to get through in a very limited time. He put down the pen and business started.

He went through the papers relating to this case, or whatever it was officially tabulated. Then turned to Burke:

"Constable Burke, I see no reason whatever for granting an exhumation order. The evidence does not justify it."

Burke expostulated. I don't know what he said, but again the magistrate examined the papers. Eventually he dismissed the whole business. I walked from the room in a daze.

On the claim matters steadily went from bad to worse. Not only were we getting no opal but Tom and Joe watched one another like cat and dog, some deep buried grudge between them.

Soon we would be compelled to sink a new shaft. It would be a hundred feet deep at least, probably hard sinking, necessitating the use of explosives. Every day two of us would work down the new shaft. The other man would work the old shaft, mullocking up. One day Tom and I would be working together sinking the new shaft; the next Black Joe and I; the third day Black Joe and Tom and so on.

I was thinking it over one morning while working at the windlass. Straight down below from where the buckets were coming up was where old Dad had died. Soon, we'd be sinking that other shaft. Then I'd be down below with Tom on top. Next day I'd be working down the old shaft while Tom was down in the new with Black Joe on top. Or else Black Joe would be below and Tom on top. What if another death – an accident occurred in Deadman's Claim!

That night at camp I said to Tom: "I'm fed up, I'm pulling out."

"Why?"

"Oh, it just doesn't seem the same since the old man died. It may be the worry, since, that's sickened me of the claim. But I'm not going to sink another shaft, I'm not going to work the claim any more. You and Black Joe can have it."

"If you are pulling out, then so am I!" declared Tom.

"Why?"

"Dunno. For the same reason as you perhaps. Anyway I'm sick of it and glad of the chance to pull out. And that's settled."

Tom and I went down the Gully to work. But we didn't last long; just drifted apart.

I scouted around the field a week seeking a place to work. There were plenty. Several months later something took me back up the Hill, past the big dump of Deadman's Claim to the very crown of the Hill. Deep shafts were scattered over the crown now, many trees had been cut down for logging and for windlass barrels. Down below the claims were chambered out into caverns, scores of thousands of pounds worth of opals had been found here since Tom and I sneaked up the Hill to sink that shaft for old Dad.

I gazed at the dumps of one worked-out claim. It had proved extraordinarily rich in opal; fabulous parcels had come out of it, and all sold in the rough. It was said that for every £100 the buyers gave for opals from this particular claim, they had made £1000. The party of elderly men who had worked the claim knew little of opals and opal-mining. To them, a bird in the hand was worth many in the bush. When by luck they bottomed on rich opal they sold the rough stones as fast as they could dig them out, wondering that any man could be so foolish as to give good money for worthless-looking stones.

That little party kept closely to themselves and invited no confidences; did not mix with the gougers; knew very little of what was going on in the field; were quite content to sell their opal day by day to the visiting buyers who so regularly called at their camp. As for the rest, they kept strictly to themselves. Down below, it was believed that they had taken all the ground out, worked it completely out in a face. Then, had quietly packed up and left the field.

Here then, was a chance. Comparative new chums. A rich claim. Surely they might have missed a little ground here or there that might still contain a few stones!

Lowering a rope, I climbed down into the darkness of the main shaft. Lit the candle, and gazed around. As the candle feebly lit up the gloom I laughed for in ghostly outline stood out a big cementy pillar of opal dirt, a huge pillar, obviously left as support for the roof. Then, when the place was chambered out the new chums daren't remove the pillar lest the mighty roof crash down. No such fear would stop me.

I didn't even glance at the pillar then; it could wait. Opal was there. What a huge cavern! As far as the little flame could reach was gloom, and ghostly piles of mullock. They'd chambered far out from the shaft in an ever widening circle, kept on gouging out the opal rock until the opal cut out. The smooth, flat roof was a huge expanse of sandstone, a bit terrifying gazing at it thus and estimating the tens of thousands of tons of solid rock hanging overhead, supported by one lone pillar.

But it wasn't supported by the pillar. That solid body of rock supported itself, otherwise it would have crushed that pillar flat as a pancake. The Ridge was wonderful standing ground, rarely indeed was a stick of timber ever necessary. But it was easy to understand why miners, not used to good holding country, had left this massive, solid block of opal dirt to support that vast roof. They must have felt glad when they cut out the last of the opal and scuttled up the shaft to safety.

I began to chip away at that pillar. Sparks flew at first, the opal dirt was a hard white cement, most opal dirt is quite soft rock. Within ten minutes the pickpoint went Click! and there under the candle-flame glowed a sparkle of orange and red. I laughed: it was always a safety valve when striking opal. The thrill and tense excitement of striking opal is wonderful.

Day by day I chipped into that pillar, only driving at the rate of a foot a day. And on opal all the time. Not rich stuff, but a nice class for all that.

Now and again I'd sit back and enjoy a smoke-oh. Then, in the cool gloom, the very silence seemed to listen. An opal mine has a silence all of its own. I know well the silence of the bush, the different silence of a Central Australian night, the unearthly silence of the Lake Eyre desert country, the wild silence of the farthest Kimberleys, the sweet, warm silence under the beautiful stars of the tropic seas, and the silence of an abandoned gold-mine. All these silences are different, strange though it may seem. And the dull silence of an opal-mine is different again. How utterly different to millions of years ago when waves and storms were roaring where now is the silence of an underground cavern!

One day, while chipping at the face, there came a "tap! tap! tap!" I listened a moment, then thinking it imagination, worked on. Presently there came again "tap! tap! tap!" I put down the pick and listened. Utter silence. What a trick of the imagination. I was about to knock off for dinner when once more came "tap! tap! tap!"

There was no doubt about it, someone was working very, very close, actually driving towards me from the other side of the pillar.

What utter nonsense! But I put down the pick and crept around the pillar. Of course there was no one there. And as far as the candle could throw light was only gloom and darkness and mullocked-up space. If I'd been working in solid ground then the soft thud of a pick blow coming this way would be audible a little distance away. But far around this pillar was worked out space. Even if it were not so, the nearest men were working far too distant for their picks to be heard. And there was no entrance from any other workings to this shaft. No, I was alone here, no one was with me.

I blew out the candle and climbed up the shaft. It was midday, time to boil the billy. Up at the surface, I climbed over the logging and stood on the dump, gazing straight across at the dump of Deadman's Claim.

Old Dad. I thought of him as I strolled down to camp. Poor old chap. After all, what use would opal have been to him, even had we struck it rich. His great faith had been that opal was there down below. And he had lived to prove his faith correct.

Next day I was working away when "tap! tap! tap!" Very softly, very distinctly. My hair slowly rose on end. I sat back, staring, waiting. Not a sound.

Every day, as gradually I worked that pillar out, those taps came again and again. Sometimes when they came I'd gently answer with a tap! tap! tap! Then listen: "tap! tap!" would come in reply, though only occasionally. It was weirdly frightening when it did come so. Who was tapping to me? To whom was I tapping? On several occasions I received the tap answers for quite a long time. He would tap, I would tap; he would tap, I would tap. But mostly the taps would come when I was deeply concentrated on prising out an opal, and their effect was more startling than if they'd been heavy blows. But for being on opal, I would not have stayed in the claim another day. Occasionally the tapping sounded as if from within the very heart of the cement pillar. But generally it came from the other side, as if someone sitting directly opposite was softly driving through the pillar towards me. Yet, there was no one there; there couldn't have been.

Day by day the pillar grew smaller. At last the day came when I chipped it all away. Only then did the tapping cease. I won several hundred pounds worth of opal from that pillar, a nice little cheque for a short holiday and to see the sisters in Sydney. Then heigh-ho to fresh fields and pastures new.

Old gold prospectors on the field were continually yarning over the camp-fires of some wild country called Cape York Peninsula, in the extreme north of North Queensland. Of the gold and the tin there, of the sandalwood and wild natives, of crocodile-infested rivers and untrodden lands. For long I'd wished to go there.

And after that holiday I went.

ETT IMPRINT has the following ION IDRIESS books in print in 2024:

Prospecting for Gold (1931)
Lasseter's Last Ride (1931)
Flynn of the Inland (1932)
The Desert Column (1932)
Men of the Jungle (1932)
Drums of Mer (1933)
Gold-Dust and Ashes (1933)
The Yellow Joss (1934)
Man Tracks (1935)
Over the Range (1937)
Forty Fathoms Deep (1937)
Madman's Island (1938)
Headhunters of the Coral Sea (1940)
Lightning Ridge (1940)
Nemarluk (1941)
Shoot to Kill (1942)
Sniping (1942)
Guerrilla Tactics (1942)
Trapping the Jap (1942)
Lurking Death (1942)
The Scout (1943)
Horrie the Wog Dog (1945)
In Crocodile Land (1946)
The Opium Smugglers (1948)
The Wild White Man of Badu (1950)
Outlaws of the Leopolds (1952)
The Red Chief (1953)
The Silver City (1956)
Coral Sea Calling (1957)
Back O' Cairns (1958)
The Wild North (1960)
Tracks of Destiny (1961)
Gouger of the Bulletin (2013)
Ion Idriess: The Last Interview (2020)
Ion Idriess Letters (2023)
Walkabout (2024)

Milton Keynes UK
Ingram Content Group UK Ltd.
UKHW010844010724
444982UK00002B/172